THE
SCYTHE
BOOK

THE SCYTHE BOOK

MOWING HAY, CUTTING WEEDS, AND HARVESTING SMALL GRAINS WITH HAND TOOLS

DAVID TRESEMER

THE PRACTICAL USE OF THE SCYTHE

PETER VIDO

STACKPOLE
BOOKS

Guilford, Connecticut

STACKPOLE
BOOKS

Published by Stackpole Books
An imprint of The Rowman & Littlefield Publishing Group, Inc.
4501 Forbes Blvd., Ste. 200
Lanham, MD 20706
www.rowman.com

Distributed by NATIONAL BOOK NETWORK

Copyright © 1981, 1982, 1996 by David Tresemer
Copyright © 2001 by Peter Vido

First Stackpole Books paperback edition 2021
Illustrations by Tara Devereux

CREDITS:
7: Chansonetta Emmons (Colby College Art Museum, Waterville, Maine);
18: Ferdinand Hodler, "Der Maher," 1910 (Aargauer Kunsthaus, Aarau, Switzerland);
19: Detail of Francis Alexander's "Ralph Wheelock's Farm," 1822 (National Gallery of Art, Washington, Gift of Edgar William and Bernice Chrysler Garbisch);
23: Detail of painting by E. B. Nebot, 1738 (Buckinghamshire County Museum, Aylesbury, Bucks);
29: Jean Francois Millet, "Death and the Woodcutter," 1859 (Ny Carlsberg Glyptotek, Copenhagen);
30: Terracotta votive mask, Attic period (Museum of Fine Arts, Boston, Catherine Page Perkins Fund);
31: Emile Bernard, "Le ble noir," 1888 (Josefowitz Collection, Lausanne);
32: Thomas Hart Benton, "Cradling Wheat," 1939 (St. Louis Art Museum, St. Louis);
All the other figures were drawn by Tara Devereux.
Quotations from *Anna Karenina* by Leo Tolstoy courtesy of Bobbs-Merrill Co., Inc.; from *The Poetry of Robert Frost* (edited by Edward Connery Lathem, 1923, 1934, © by Holt, Rinehart and Winston, copyright 1951, © 1962 by Robert Frost), by permission of Holt, Rinehart and Winston, Publishers.

British Library Cataloguing in Publication Information available

Library of Congress Cataloging-in-Publication Data available

ISBN 978-0-8117-3979-5 (paper : alk. paper)
ISBN 978-0-8117-6986-0 (electronic)

♾™ The paper used in this publication meets the minimum requirements of American National Standard for Information Sciences—Permanence of Paper for Printed Library Materials, ANSI/NISO Z39.48–1992.

Preface

Twenty years ago, I learned the right use of the scythe from Eliot Coleman. I then went on to design an American version of the European snath, imported the best of the European blades, and put together an integrated system of tools to sharpen and care for this wonderful means of cutting grasses, grains, and brambles. My company was called By Hand and Foot, Ltd.: Tools Dependent on Human Energy, which later became Green River Tools, and we sold many garden and farm tools by mail order. Ten years ago, Green River Tools was sold, and the new owners dropped the scythe line. Other companies who copied the snath design and other parts of the system deleted them from their offerings. For a long time, those who wrote me had no idea where to obtain good equipment, and I was hard pressed about what to tell them.

Now good equipment has become available in catalogs, a combination of imported and homemade. Even if you don't have the very best components, once you understand the scythe's principles, you can work with any part of the system to refine it—bending and sharpening the cheaper, duller blades, rasping down the cheaper, heavier handles—and bring back the thrill of a body exhilarated by the outdoors while the blade whispers through the plants.

Further work on improving this tool, and understanding how it works so well, continues. In his book, *Living Energies* (Bath, England: Gateway Books, 1996), Callum Coats offers an analysis of the scythe's superiority to the wounding of mechanical cutters: "The scythe delivers a slicing cut, thereby keeping the wound area to a minimum. Due to the energies inhering in the blade and implanted there by hammer-sharpening, the imparted ionization quickly draws the elements in the material surrounding the wound together and rapidly seals it." (p. 261) In a series of articles for *Small Farmer's Journal* (from Sisters, Oregon, especially the issues of spring and summer of 1995), the well-read and antitechnological Canadian Peter Vido shares his knowledge about how to work this amazing tool. And the professional archaeological journal *Tools and Tillage* (from the National Museum of Denmark, at Lyngby) publishes pieces about these tools and their ancestors, including experiments with old tools that reveal their best use. All of these recent developments have so far supported the tenor and conclusions of *The Scythe Book* and extend its invitation to use the scythe.

In this age when most wonder how you can accomplish anything without the aid of electricity or gasoline, we must reassert that a hand tool finely made has greater versatility, attains comparable efficiency even in large areas, and even makes for good exercise because the tool fits the dynamics of the body. Most importantly, a finely made tool does not interfere with personal communion with the ever-amazing natural world. Without the noise and bother of the big machines, and with the heightened senses of a body electrified by activity, we open ourselves to the profound teachings of *Natura*.

The companies come and go. The meadow beckons forever. Good mowing to you.

—D. T., 1996

Sources

True high quality European-style scythes are available from the following companies:

Lehman Hardware and Appliances, PO Box 41, Kidron, OH 44636; 888-438-5346, lehmans.com

The Marugg Co., PO Box 1418, Tracy City, TN 37387; 931-592-5042; themarugg company.com

Scythe Supply, 496 Shore Road, Perry, ME 04667; 207-853-4750, scythesupply.com

Acknowledgments are due to Eliot Coleman for introducing me to the European scythe; to Tommy Thompson for introducing me to its best use; and to Castle Freeman and Susan Tresemer for helping me to write about it; also to Keith Squires, Stephen Bourne, and Brian Gabree for assisting in time trials; to Tommy Thompson, Rachel Hillel, and Peter Payne for reading portions of the manuscript and offering suggestions; to Bob Anderson and Irving Perkins for carefully guiding the production of the printed work.

Contents

The Scythe Must Dance

Introduction

Years ago I bought a scythe at the local hardware store. I bought it to keep the dandelions, milkweed, lamb's quarters, couchgrass, and so on from going to seed. There were too many to pull up by hand, and too few to justify the hiring of a farmer's mowing machine. I had watched powerlessly as their flowering and seeding progressed. I wielded my scythe a few hours, then hung it up. It was awkward, it left me sore, and the grasses laughed at my efforts by bending over and bobbing back up after the blade had passed. I concluded that our ancestors were made of stouter stuff than I am!

I learned later that the scythe I had used was the traditional "American" type, having a bent ash snath and a narrow, hard steel blade. Five years ago, I was introduced by Eliot Coleman at the Small Farm Research Association in Harborside, Maine, to the "Austrian" style scythe, which has a light blade and snath. My first use of this scythe was in happy contrast to my earlier labors with the "American" scythe. The experience was marked by the same observations anyone can make on discovering a good tool:

- *It fits:* The scythe complements the contours and dynamics of my body firmly and comfortably, and feels like a well-designed extension of my hands and feet.
- *It doesn't hurt:* There is no excessive strain on any one part of my body as I use the scythe, and I can use it hours at a time without abusing myself;
- *It works:* We work together; my energy is efficiently translated into the desired effect without waste of effort. I can tell I have found a good tool when I hesitate to let it go. With the Austrian-style scythe, I went on mowing and mowing.

I use a scythe to mow a few acres of hay which I use to mulch my gardens and to feed the small livestock population I keep over the winter. I also harvest with the scythe, a variety of small grains for bread and pancakes. With a very modest expenditure of time and effort, I am thus able to raise my own hay and grain. With the scythe I keep the marginal areas of my property free of weeds going to seed and have thus reduced the weeding problem in my gardens by at least the time taken in the scything. I have found that a scythe can perform a moderately sized task in the same amount of time it would take to fetch, attach, adapt, and repair a mechanical substitute. Maintenance of the machine means money spent; maintenance of the human body means health gained. It is not impossibly strenuous or difficult to wield a scythe; both my experience of mowing and the tool itself are resources of enduring value.

Before the widespread use of mowing and harvesting machines, commencing in the latter part of the nineteenth century, the scythe was the main tool responsible for harvesting hay and grain, as well as for keeping trim and neat the estates of rich lords and modest freemen. As with any common tool, little of the early writing about the scythe explains its use, which was general knowledge. Beginning in the early part of the twentieth century, however, increased interest in rural life has led to the acceptance of a form of autobiography, summarized as "my experiences back on the farm," in which the authors, recording the recollections of old country people, often include some instructions in forgotten practices from the old-timers whom they encountered. Some of these writers have become good mowers themselves and have passed on what they learned.

From these autobiographies, from old agricultural journals, from the evidence of drawings, paintings, and sculpture, and from the work of anthropologists and historians, we can piece together a fairly accurate picture of how the scythe was designed and used. There are many small variations depending on period and region. I shall not document them all. Suffice it to say that for every "rule" or tradition I have for scythe design and technique, somebody someplace else did it a little differently and got the grass cut.

There is a certain romance in using an ancient tool the old way, experiencing what our ancestors must have experienced. However, a tool must still be efficient today. That is, romance aside, it must make such an effective use of a person's time and energy that it is competitive with other means for accomplishing the task at hand. A chipped stone knife, while representing an immense technological leap for its time, does not meet this criterion. A sickle for harvesting grain does meet this criterion for small areas. A good scythe meets the criterion of practical efficiency for larger areas, and for a wider variety of tasks.

University extension agents argue persuasively that a farmer cannot afford the time to mow: he is forced by economics to sit all day atop his tractor attaching one implement after another to produce enough to make the payments on the loan for the equipment. I do not offer the scythe and its related tools as an alternative to this scale of agriculture. Yet, even here the scythe has a place, trimming the edges where the machine cannot go.

I do not use a scythe in order to make enough money to support a livelihood, but rather simply to support a livelihood. Not only is the product of my labor—hay, mulch, grain—superior to any I could obtain elsewhere in this chemical age. In addition, I deeply enjoy the experiences of contributing to my own sustenance and of relating intimately to the earth.

THIS BOOK

The first section of this manual covers the parts of the scythe in more detail than the casual mower probably wants, I began this writing expecting to produce a brief manual on mowing technique, but found the details of manufacture and tradition and so forth to be so important to my mowing that I took the research as far as it could go, and the manual became a small book. I asked one American writer who had written that he was "one of the very few true authorities on the scythe" what resources he used for his technique. He exclaimed, "Are you kidding? Just go outside and swing the thing!" That was all he would say about how to use the scythe. I am, however, convinced that with the right tool, the casual mower will soon become a serious mower, at which time the detail I provide here will be more interesting. In any case, though, reading the sections on sharpening and mowing technique is all that is required for stepping out to encounter the legions of dandelions. Sections follow which set out the minimum considerations necessary to mow hay, cut weeds, and harvest grains.

The Parts of the Scythe

THE BLADE

THERE IS A STORY TOLD IN LEGRAND CANNON'S *LOOK TO THE MOUNTAIN* IN which the master blacksmith makes a scythe blade of Damascus steel for a young mower in New Hampshire in 1769.[1] In the high heat of his forge he pounds bars of iron and steel together, folding them over each other again and again. After tempering, the effect of the mingled metals in the blade is much like fudge ripple. The iron gives the blade flexibility as an antidote to the steel's brittleness; the steel holds the edge razor and gives the enduring shape of the blade. The price? Twenty-one cords of rock maple, cut, split, and stacked. The master smith assures the young man that there will be at least one blow of the hammer for every strike of the ax.

Today the common hardware-store scythe blade is stamped in a powerful drop-forge press and its edge is ground sharp on a high-speed grinding wheel. In contrast, the edge of the sharper and more delicate European- or Austrian-style blade is hammered to the correct thinness, in much the same way it has been since the fifteenth century, when the scythesmith guilds were established in the mountain valleys of Austria. The advent of water-driven trip hammers, and, now, of small but powerful hand-guided air hammers, have made fine scythe blades cost less than twenty-one cords of split hardwood, but the hand forged Austrian blade is still more expensive than the stamped blade. The process at its best involves a set sequence of twenty-six steps, pictured in Figure 3. Although little change is apparent between steps 14 and 20, here is where the most important tempering and shaping take place. The blade is heated in a furnace, cooled in warm oil, sandblasted, and diagonal-hammered, to give the optimum curve and thickness to the back and edge.

The common hardware store blade is usually called the "American" type, the better blade the "Austrian" type. In fact, scythe blades have not been manufactured in this

Figure 1

The parts of the scythe

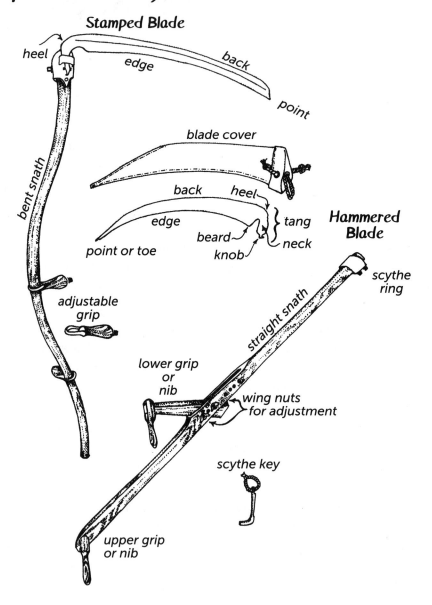

Stamped Blade

heel

edge

back

point

blade cover

back   heel

edge

point or toe   beard   knob   neck   tang

Hammered Blade

bent snath

adjustable grip

straight snath

scythe ring

lower grip or nib

wing nuts for adjustment

scythe key

upper grip or nib

Figure 2

Stamped and hammered blades

Stamped Blade   **Hammered Blade**

face view

edge view

edge & tang

end view

edge   tang   25°

hafting

3¼"

country for nearly twenty years, and, when they were, some were of fine quality. The "American" pattern blades are made in Europe by the same companies who produce the finer blades; they freely admit that the stamped blade is the bottom of their line.

A fine scythe blade must be light in weight, and tempered to carry an edge that will cut tissue paper without tearing. Yet it must be soft enough to dent instead of break if a stone or stump is accidentally hit, and capable of having its edge restored through hammering[1] with a light hammer.

The problem is to find a steel of the right consistency to achieve this wonderful balance of hard and soft. In England, "Crown" blades were once made with an edge of hammered iron sandwiched between two pieces of mild steel to hold the overall shape.

Figure 3

Twenty-six steps in the manufacture of the hammered blade

Blades are still made in England in which the edge steel is rivetted to a stronger back of "mild" steel.[2] The New Hampshire blacksmith of 1769 had to mix the iron and low-grade steel to get the right effect. In one tradition, mowers hung their scythes in a tree for the winter to rust the iron out, thus—according to the tradition—raising the percent of carbon, and the hardness blade. The carbon content of the "American" blade is already fairly high, and in fact makes the blade too hard; the impact of accidental strains from stones and stumps leads not to dents but to microscopic cracks which will in time cause the blade to break.

The shapes of the two types of blades differ (Figure 2). The stamped blade is expectedly flat. The hammered blade is curved in every dimension to optimize the relationship of the body to the task of cutting. The "beard" and heel are distinguished in the hammered blade because of their distance from each other; in the stamped blade, that whole end is called the heel. The (queue or tail in French) is the very strong projection from the back rib which attaches to the snath; the parts of the tang are the heel, neck, and knob.

The manufacture of the hammered blade shown in Figure 3 begins with a high-quality steel, and, by working it hot and cold, aligns the molecules and strain-hardens the steel. Exact requirements for these blades make it clear why over half of the blades which begin the twenty-six stages of manufacture are rejected along the way. Omitting any of the manufacturing steps makes other brands of blade less expensive, but only at the loss of the desirable characteristics of the finest blades.

The better blades are stamped on the tang with the trademarks of the manufacturers (Figure 4) which can be checked in the directories of Franz Schrokenfux and Josef Zeitlinger. These references match the trademarks with the methods of manufacture of the particular company as well as the genealogy of the master, leading to the influences of the Ottoman Empire in the fifteenth century. In recognition of these Turkish origins one brand has the name *Turkensensen*, or "Turkish scythes."

Several models of blade have a series of a hundred hammered dots that look like a ball bouncing across the edge of the blade stopped in flight by a stroboscopic light. The dots are both a fanciful ornament and a testimony to the skill that guides the hammer.

When the scythe was the predominant tool for many agricultural tasks, blades were made in different lengths, from very short blades for careful work around vines and trees, to very long blades—over four feet in some areas—for less exacting tasks such as mowing expansive estate lawns. The length now used most frequently on European farms for the

Figure 4

Some blade trademarks

enlarged detail

widest variety of tasks is 27½ inches (70 centimeters). Bush or brush scythe blades are as short as fifteen inches and often thicker to withstand repeated blows against young saplings. The grass blade of 70 centimeters is adequate for all grasses, weeds, briars, and one-year old trees.

BLADE COVER

You cannot cut yourself with a scythe when you are holding the grips. In transit to or from work, however, the blade can be dangerous to oneself and others. A good way to prevent cutting people and nicking the edge is to slip on a sheath that completely covers the blade and tie it securely around the heel end. The narrow strips of plastic sometimes taped to the blade to protect the edge from the parcel service should be discarded after delivery, because they are unreliable for repeated use.

At the commencement of cutting, remove the cover and put it folded in your pocket. At the end of cutting, wipe off the grass and dirt from the blade, and put the cover back on. Secure it with a simple overhand knot behind the heel of the blade. At the end the season, the scythe should be stored high up out of children's reach. And remember that the cover does not protect the legs of bystanders during mowing, or the hands of the sharpener.

When scythes were used more widely, the blades were covered with a bag or custom-made sheath when not in use. In one area in England where scythe covers were not used, the tradition was to march en masse to the field with the blade in the air, the right hand holding the top grip and the lower grip thrust firmly under the right armpit. This regimentation prevented mishaps.[3] In every area, during the break for the midday meal, the scythe blades were laid in the uncut grass or grain, into which no man, woman, or child would dare stray, thus protecting blade and flesh. In modern times, these traditions are not known to all, and the blade cover gives a necessary margin of safety.

SNATH

A great variety of *sneads* or *sneds* or snaths have been used with the same long, curved scythe blade. One English tradition used only naturally curving willow saplings, while in East Europe an offset handle was mortised into a straight pole, usually of ash or hickory. A traditional Scottish design had a long offset handle which met the snath almost at the scythe ring. The scythe is one of the few asymmetrical hand tools; with it the hands do not perform opposite or interchangeable tasks.[4]

The grips (or nibs or nogs) have been attached to the snath in many different ways, sometimes pointing out in the same direction as the blade, sometimes in the opposite direction, and occasionally mixed. Adjustable grips, made of a wire hoop around the snath tightened by twisting the wooden grip, have been found in museum collections pointing in nearly every direction. In some traditions only the lower nib is used; in others the upper part of the snath is cradled in the crook of the left elbow.[5] These countless variations can be expected from any tool which is locally made.

The typical sweeping curved snath found in American hardware stores is made with bent swamp ash from the Southern United States. The poles are steamed, bent to the desired shape, and dried. The problem with this method is that the amount of wood necessary to keep a permanent bend increases the weight of the snath. Old-timers tell of going to the farm supply store to buy a snath and taking half a day to try out each one of the snaths in stock, all of them slightly different. To reduce this variability in shape, snaths are now made even heavier, so each one turned out holds an identical pattern. Several samples weighed in 1980 averaged 4¾ pounds. With the heavier stamped blade, this added up to 6½ pounds for the bent ash snath with blade, versus 3½ pounds for the hammered blade and straight snath. Consider what this difference in weight will mean after the approximately ten thousand strokes required to cut an acre of grass.

It seems to me that the curving "American" snath, with handles which rotate completely around to point in any direction, tries to bring the angles of the hammered blade (Figure 2) to the simpler stamped blade. I prefer the simpler snath and the better blade. Sometimes an old bent snath will be very light and well-balanced, the primary disadvantages being the wood is dried out and fragile, and the adjustable nibs are permanently loose. Also, some old scythe rings were so small that only certain very narrow tangs could fit through them.

The weight of the heavy wooden snath makes more serious the tendency of the American-style scythe to put the right wrist in a cramped position. This position in which the little finger is bent back toward the wrist ("ulnar deviation") reduces the capabilities of the wrist and may lead to strain in the tendons.[6]

Lighter snaths in the American style are occasionally available made from aluminum, but I confess to a suspicion of aluminum used in tools under stress. Unlike steel, aluminum gets weaker and weaker with use—there is no "fatigue limit," meaning no limit to weakening of the metal from normal use.[7] Eventually it breaks. Even a little bend can upset the angle of the snath to the blade.

After three years of experimenting with different snaths from several different countries, I decided that a slightly curved tubular steel snath made in Europe was the best available for the average user. It is light but strong; the blade is set at the correct angle to the snath; the lower grip nearest the blade can be adjusted though the upper grip is fixed. Every adjustable grip eventually becomes loose and must be shimmed with a bit of hardwood or leather. The knob of the tang fits into a slot at the end of the snath and is secured by two screws in the scythe ring tightened with a special key for that purpose. The key can be stored in the very end of the blade cover or looped around the belt; if the key is lost, the ¼-inch-square adaptor from a socket wrench set fits the screws of the scythe ring well. The snath needs very little care except a cursory wipe at the end of the day.

I have also developed a straight wooden snath made of airdried Vermont ash. Its design was synthesized from several different models currently used in Europe. It has the additional feature of a mortised channel for the lower grip, permitting adjustments to different body sizes. It is light, strong, and permits use while standing erect. Like all wooden tools, however, a wooden snath requires extra care: if it gets too wet, it can warp slightly and change the angle between snath and blade. Then it needs adjustment. It should be lightly oiled at the beginning of the season—one part linseed oil to one part turpentine is a penetrating and clean solution. The section between the lower grip and the blade should be oiled again after each extended use of the scythe in wet grass. The choice to varnish instead of oil must be made at the very beginning, and the snath requires four or five coats of varnish to be absolutely certain that every pore in the wood has been filled.

Without this little extra care, the wooden snath can become a less useful tool than the tubular steel snath. Therefore, this snath is best for those with some prior experience of mowing on which to base their decisions for adjustments. The advantage of the wooden snath is in the user's ability to perfect it as a tool uniquely fitted to him- or herself, a graceful tool closely complementing the individual's body proportions, rhythm, and style. It is a grander snath than any other and permits full use of the blade's virtues in a balanced swing. The flexibility of wood has a nice feel as an intermediary between the flesh of the hand and the metal of the blade.

Several rules for the placement of blade in relationship to snath, and for the adjustment of the grips along the snath, recur repeatedly in the old accounts of the art of the scythe:

1. If the heel of the blade is set next to the heel of the right foot, the first grip should come to a point just below the right hip bone (Figure 6).
2. When the scythe is held in this position, the left foot should swing out along the blade following its curvatures. Continuing this arc with the left foot gives a good idea of the shape of the longer blades once used for mowing lawns. To obtain the angle of the blade and snath (the "hafting angle") more accurately, set the snath against a building with the upper grip resting on the ground and the snath in the "twelve o'clock" position (Figure 5). Mark the point where the "beard" of the blade (the edge nearest the snath) touches the wall. Now pivot the snath to the right on its upper grip so the snath is at about two o'clock, and the tip of the blade is in a vertical line with the mark on the wall. The tip should be three fingers *below* the first mark. This method can be adapted to setting in the field by using marks on the ground.
3. The two grips should be as far apart as the length between the elbow and the fingertips (the Mesopotamian cubit). This puts the top grip at the shoulder (Figure 6). This measurement can vary according to conditions, in easy cutting becoming as short as four hands (a hand being about four inches).
4. The scythe should balance with the blade parallel to the ground on one finger held in the middle of the lower grip.

The scythe blade is twisted, or cocked, with respect to its tang. When scythe blades were sent out with no angle to the tang, and the blade was attached with a simple ring and wedges, the fitting required a careful bending in a vise until the tang is rotated to an angle with the blade of about twenty-five degrees (Figure 2, end view).

The basic rule here is that the scythe can be dragged on its rounded back without engaging the edge in the ground, but with the edge no more than ¼-inch above the ground. This is how the lawn is mown, and if you can mow the lawn, then the blade is properly cocked and hafted.

Figure 5

Correcting the angle of haft

mark height
of beard

3 fingers
difference

mark height
of point

with hardwood shim to
hold blade in rough cutting

Figure 6

Adjusting the nibs

fingertip
to elbow distance

lower nib
at hip height

These rules should give a general idea of relationships between the parts of the scythe, but different designs of scythe, different people, and different conditions of cutting sometimes require modifications.

The straight snath of air-dried ash can be fitted to any size of adult in a way not possible with any other available snath. The upper grip is fixed. The lower grip is adjustable within a mortised slot and is secured by two carriage bolts and wing nuts (Figure 1). This grip should be set at a cubit's length from the top grip, or a little further. The snath is quite long so that the heel can be cut by the owner to a different length if necessary; the lower grip should come just below the hip bone when the scythe is resting on the heel of the blade. A hole ⁵⁄₁₆-inch wide and ½-inch deep must be drilled in the center of the snath 3¼ inches from the end into which the knob of the tang fits (Figure 2). The blade is hafted as in number 2 above and held in place with the set screws of the scythe ring.

You can experiment with leaving the snath a little longer than suggested above, but if it feels too unwieldy, and the blade seems to be pulling you over when you cut, then trim the snath down to size. The scythe ring keeps the blade tightly in place, but if there is a problem with the tang working on the drilled hole, then a piece of metal (such as roof flashing or eighteen gauge sheet steel) with a ⁵⁄₁₆-inch hole in its center can be tacked onto the snath for strength.

A smaller size of snath for small people and for boys and girls between six and fifteen is also manufactured, and is fitted in the same way. A shorter blade (40 centimeters or 15¾ inch) is best for children. Such a tool is used in Switzerland, not only for young people but also when mowing the forty-five-degree slopes in the mountains. In areas where snaths and blades were by tradition heavy and long, scythes were not used by children for harvesting; in order to use the labor of the children the sickle was the traditional tool. An adjustable snath, however, gives smaller people the opportunity to stand erect when mowing; and learn the enjoyment of working together with their friends. I have used Tom Sawyer's technique of getting the fence whitewashed with adult mowers and child mowers. I find that, once the cultural aversion to physical work is overcome, we have the good time of shared work.

Sharpening

The faults (which the unlearned man) hath he will learn how to hide and color them, but not much to amend them; like an ill mower, that mows on still, and never whets his scythe.

— Francis Bacon, *The Advancement of Learning*, 1605

He, therefore, that spends his whole time in recreation, is ever whetting, never mowing.

— Bishop Joseph Hall, *Occasional Meditations*, 1633

WITH THE "AMERICAN" BLADE OF HARD STAMPED STEEL, THE TRADITIONAL method of sharpening was on a large circular grindstone (Figure 7). Thus Robert Frost could write about turning the crank as a boy while an older man held the blade at just the right angle to the rotating stone. Frost complained about the length of time it took and the wear on his arms; violent emotions arose toward the end.

> Once when the grindstone almost jumped its bearing
> It looked as if he might be badly thrown
> And wounded on his blade. So far from caring,
> I laughed inside, and only cranked the faster,
> (It ran as if it wasn't greased but glued);
> I'd welcome any moderate disaster
> That might be calculated to postpone
> What evidently nothing could conclude.

Figure 7

The grindstone (E. Portland, Maine, 1910).

Today the handyman sharpens this type of blade with a small grindstone rotated in an electric drill, followed by a file, followed by a coarse Carborundum or Crystolon "scythe stone."

Yet old accounts tell of taking a strickle to the fields—that is, a stick scored with a saw, which was covered with grease and rolled in sharp sand. The *strickle* was used to touch up the edge of the blade while the mower was working in the hayfield; but this blade was made of a finer more malleable steel than the hard blade that Frost helped to sharpen. The grindstone would cause excessive wear on it: Worse, the grinding would overheat the area of the blade in contact with the stone, leading to loss of temper; after grinding, the softer, finer blade would not hold its edge as before.

The method of sharpening appropriate to the Austrian-pattern blade involves two steps: 1) *peening*: hammering the blade upon a small anvil; and 2) *whetting*: the final honing with a whetstone.

PEENING

Peening is practiced on American and European farms today in a way unchanged from the way it is illustrated in Brueghel's sixteenth-century painting, "Haymaking." A small anvil is set into the earth or into a bench, and the blade is lightly hammered with a bar peen or cross-peen hammer (Figure 8). The lightest cross-peen hammer made in this country is an expensive blacksmith's hammer weighing over two and a half pounds; the imported hammer weighs about one and a half pounds.

The edge of the blade rests perfectly flat on the anvil. The hammer blows begin about ⅜-inch away from the edge, and proceed to the edge. It is easier to achieve this stroke with several fast and light taps in the direction of the sets of four blows shown Figure 9. The bar peen strengthens the steel along the edge; either the hammer or the anvil or both must have this bar. Two flat faces or a common ball-peen hammer would not properly align the steel molecules along the edge. The bottom of the blade must be against the tool with the flat face, so that when using a flat hammer and a bar peen anvil, the blade must be topside down.

To keep the blade flat on the anvil, you may balance the snath over your right shoulder, or another person can hold the top grip at the right height while you hold the blade with one hand and the hammer with the other. The top grip can also be tied to a string from a branch or rafter, as diagrammed in Figure 10; the person then can sit on either side

Figure 8

Peening: hammer, blade, and anvil

of the blade. The whole process of peening takes five minutes, and should be done every twelve hours of use, or as necessary to straighten out an accidental dent in the blade.

The cold-hammering of the tempered blade draws out the steel a tiny bit with each blow, to maintain the ideal thickness. Most people do not believe that it is possible to shape steel without a blacksmith's forge. With the hard "American" blade, it probably isn't. However, the crystalline structure of high quality tempered steel can be molded. In the molecular realignment of cold work, the steel becomes strain-hardened without losing its ability to dent under severe stress (ductility). However, if the pounding-out is overdone and leaves an edge too thin and brittle, the blade will break at the weaker points. The object in peening is to maintain the bevel and thickness of the original blade, as it was crafted by the master smiths. A rule of thumb is to hold the blade as in Figure 8 and push up on the very edge with your thumbnail. If the edge can be slightly moved by this pressure, then it is the right thickness. If it does not move at all, then it needs more peening. If it moves but does not spring back into place, then you have been hammering for much more than five minutes!

Many claim they could not possibly hammer the blade evenly along its length, and that such hammering would soon use up all of the blade. Not so. I have seen several blades used every season for twenty to thirty years, and they are from ¼- to ½-inch narrower than their original width. The edges, no longer perfectly straight, have become a little wavy, showing that the wear and/or the sharpening has been uneven. However, this is acceptable for a scythe, and does not impair its performance. As Adrian Bell said in *Apple Acre*, "His scythe edge is as wavy as the sea; every nick is a history of his labor."

If by accident the blade strikes a stone or stump hidden by the grass, a section of the edge may be blunted, or bent over, or it may even show a tear up to ⅛-inch long. The risk of such damage should be avoided, in the first place, by not demanding the scythe to do the job of an ax or billhook. These high-quality scythe blades will cut year-old trees and brambles, but not tree trunks the diameter of your finger or larger. However, accidents with stones, fences, and trees do happen, and peening is the only way to straighten out a dent. The hammer and anvil are made light-weight with holes through which strings can be attached so they may be carried easily to the place of work for such emergencies. By repeatedly tapping around the edges of a dent, you bring the blade back to its original shape. And, if you're smart, you take the opportunity to remove or clearly mark the obstruction.

Figure 9

The sequence of hammer blows in peening

Figure 10

A simple arrangement for holding the scythe when peening

When staring down at a large dent, I must remind myself that it is better to dent the blade than not to. The dent can be mended; the blade that shows no evidence of stress has stored the damage in tiny cracks which will shorten its life.[8]

WHETTING

Anew with merry stave and jest
The shrieking hone we ply.

—WILLIAM ALLINGHAM, *THE MOWERS*, 1865

Whetting with a stone gives the blade its even razor-sharpness, and is repeated at intervals during the work. You mow once around the hayfield, or let us say you mow fifteen minutes' worth, then stop for thirty seconds or a minute to whet the blade. Seen from the tradition of the hard stamped blade, which never got as sharp as the hammered blade, and only slowly lost the sharpness it had, this pause is thought lazy. However, the higher-quality blade well repays this periodic attention to its sharpness with a much more thorough cut requiring less effort. The routine sharpening is not time wasted, since the mower starts afresh with a better tool. Using a dull blade encourages more violent swings of the scythe to do the cutting—this is tiring and potentially dangerous. The two quotes at the beginning of the chapter show a conflict in attitude about how much time should be spent sharpening; but I prefer to sharpen as often as I can break the trance of mowing. This may be every fifteen minutes.

Historically, whetstones were cut from natural outcrops of the right sort of rock. Since the advent in 1906 of mass-produced and therefore cheaper synthetic abrasives, natural stones have been used less. The hardware store "scythe stone" is usually made of silicone carbide or aluminum oxide from Canada bonded with resin and baked at 2500 degrees Fahrenheit. These common stones are too coarse for the blade of finer steel. European stones of silicone carbide have a finer grit and are more appropriate to use following the peening of the hammered blade. Natural stones are of a much finer grit still; when used for the very final honing, they impart a sort of perfection to the edge that may not be noticed by the beginner but is greatly appreciated by the mower with a little experience. To use the metaphor of sandpaper grit, the common stone is like 50 grit; the European synthetic stone, 120 grit; and the natural stone, 320 grit.

The beginner is advised to proceed slowly with whetting, wearing a glove on the hand holding the stone. The point of the scythe blade can be set firmly against a tree or post or into the ground for stability. The scythe can also be held upright with the left hand, the left foot stepping on the top grip. Wipe the blade clean with a rag or some cut grass. Hold the stone against the blade so that across the back side of the blade, the angle of stone to blade is zero degrees. You should see the first ¼-inch of the edge brighten with the whetting. The bevel of the edge is dictated by the sharpening of the front of the blade. Here the stone is guided by the rib which stands out along the back of the blade. Whet slowly; the correct angle is far more important than rapidity of strokes.

As in the drawing (Figure 12), begin with the base of the stone at the base of the blade and finish with the point of the stone at the point of the blade; from heel-and-heel to toe-and-toe, or base-and-beard to point-and-point. The reason for this sequence is that tiny serrations or sawteeth are built up pointing in the direction of the blade's point. These help the cutting of the grass. If I were sharpening a shaving razor, I would begin sharpening at toe-and-toe and hone *back*, toward the blade, to eliminate the very small burr of excess metal that occasionally builds up beyond the blade's edge. For cutting grass, this burr does not impair the cutting, and soon breaks off. In any case, it is less dangerous to slide the whetstone and the hand away from the edge rather than toward it.

One tradition recommends that the whetter stand with the light falling over the left shoulder to better see how the stone is bearing on the edge. Some say that during whetting the blade should point either north or south. In fact there are interactions between the earth's magnetic field and the alignment of steel molecules in the edge, but they are very subtle. (Boat-builders traditionally lay the keels of the boats they are constructing in a north-south direction, presumably for the same reason.)

The whetstone-holder (see Figure 11) protects the brittle stone in transit. It also holds a little water (mixed with vinegar by the Pennsylvania Dutch) so that the stone can be cleaned of the microscopic pieces of metal that clog its pores during whetting. Holders of old were made of cow's horn or wood ornately turned and individualized. But cow's horn is rare and wood cracks, and they have given way to hard plastic holders with a belt clip for carrying. The stone should be jiggled in the water every fifteen seconds during whetting to clean the pores.

At one time, much discussion was devoted to the virtues of the different slates and sarsens of every color—blue, green with red bands, brown, grey, white—from

Figure 11

Whetstones and holders

wooden
holder

quarried
stone
fine grit

European
silicon carbide
medium grit

my grandfather's
stone

side   front
view   view

plastic holder
with
belt clip

Figure 12

Whetting

Figure 13

Cleaning the blade

the different quarries local to every region. In his book, *Man the Tool-maker*, Kenneth Oakley shows a piece of sarsens (sandstone hardened by secondary silica intrusion) used to sharpen primitive axes thousands of years ago. Particularly good paving stones and monuments were once popular as stationary whetstones.[9] I have a piece of slate from the nearby Whetstone Brook and a whetstone of slate from my grandfather worn down from years of use, and have found that they do impart a very sweet edge. The final touch comes from an English tradition which I have never heard mentioned in this country, though I have tried it and been satisfied: honing with the dried shelf fungus of the sort that grows on dead birch trees (*Polyporus betulinus*), which perfectly fits the hand.

In some places the strickle was used instead of a stone, and a whole new set of tools was necessary for the sharpening process. A saw or a special iron strickle-pricker was used to make small "delfs" (holes) to take the tallow or grease; extra grease was carried to the fields in special pouches or horns; sharp sand or emery powder was found and carried to the field in little bags; snaths were sometimes made seven feet long to accommodate a strickle-holder.[10] Keeping the grease cool enough not to run off, and the sand sharp enough, and so forth—these are lost arts. I am content with the sharpness I get from my two grades of stone.

The final step in keeping a sharp edge is wiping the blade after each day's use, shown in Figure 13, preventing rust and improving the mirror-like finish. The mower Damon's blade was such that in it, "I see my Picture done, As in a crescent Moon the Sun."[11]

I do not loan my scythe. I reap the benefits of a carefully tended edge, and am reminded of my responsibility by a dull edge. Each time I peen and whet the blade, I remake the tool as my own; in time it becomes a mirror of my attitude toward work. A Finnish tale describes a tired mower who exclaimed "May the devil whet my scythe!" The devil did so, leaving only the snath.

Mowing Technique

O sound to rout the brood of cares,
The sweep of scythe in morning dew
                    —ALFRED, LORD TENNYSON, *IN MEMORIAM LXXXIX*, 1833

MOWING SHOULD BE COMFORTABLE, NOT TOO STRENUOUS, NOT—IMMEDIATELY—
tiring. If it is exhausting, it is wrongly done. Dress for mowing, therefore, should also be
comfortable. The best includes loose-fitting pants and shirt. The shoes should be light;
the feet are not threatened, and mowing was often done barefoot. A sun hat is a good
idea, and is always worn by the European mower. A scarf may be useful to warm the
neck in the early morning, and to mop the brow in the afternoon.

A HELPFUL EXERCISE

An exercise to begin each mowing session helps remind me of the shape of the movement (Figure 14). I stand erect, neither slumping nor rigidly at attention, and consider
gravity as the fundamental force of the earth, which acts upon my entire body at every
moment. Gravity is an acceleration, as opposed to a steady velocity: an ever-new force
that pulls me downward faster and faster. I respond with an acceleration upward, in
opposition, also renewed every moment, an anti-gravity. I prefer the terms *energy* or
spring to *anti-gravity* for denoting this counter movement; it is not an anti-earth force,
or a rejection. The earth is the entity from which we spring. We are the earth's *offspring*.
The other connotations of *spring* are all correct. Neither pulled back to merge with the
source, nor flying up uncontrollably in escape, I stand erect, firmly grounded yet springing toward heaven. The stance itself is an act of creativity. Although this interaction

Figure 14

Twisting exercise

between myself and gravity always goes on, it is usually unconscious. Mowing reminds me of the relationship, as it restores the balance of gravity and spring.

In the middle of these two accelerating flows of gravity and spring, I am relaxed; my shoulders are not hoisted up, my neck is not stiff, and I can mentally travel from one part of my body to another releasing unnecessary tension. Then I begin to twist my torso from one side to another, to face my right, then my left, letting my arms swing free. My right hand flops against my left thigh, and a full second later my left hand flops against my right thigh. There is twisting movement in my ankles, knees, hips, and all along my spine. The exercise serves to remind me of three things:

1. Mowing is a twisting movement, from the right side of my body around the front to the left side;
2. My arms and shoulders need not be tense, but may be relaxed; and
3. My whole body is involved in the twisting motion. The human body can be seen as two coils twisting diagonally in opposite directions up through the body in dynamic equilibrium. In mowing, this dynamic equilibrium becomes a rhythm between one side and the other.

The energy that issues in the mowing stroke is stored in our tendons and muscles as we mow. Tendons are superior to the yew wood used in hunting bows for storing energy under stress. They are even superior to modern spring steel. In Greek and Roman times animal tendons were wound in a spiral to power very effective catapults that could send a 360-pound stone a quarter mile.[12] When I am at one extreme of a twist in the body, my tendons are storing energy, as a wound rubber band does. This energy is released to aid the contraction of muscles when I twist to the other side. The best stroke uses all the tendons in all the joints.

WITH THE SCYTHE

Moving from the exercise to mowing, I let my right arm hang loosely to the side. This is where I had it at the beginning of the scythe stroke. The point of the blade is in line with both feet (Figure 15, beginning). My left elbow is cocked enough to set the blade at the desired height.

Figure 15

Beginning and end of the stroke

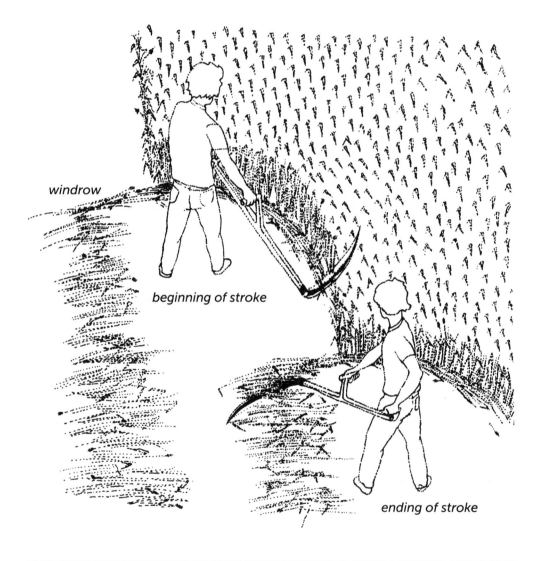

windrow

beginning of stroke

ending of stroke

I stand upright. The stereotypical picture of the man with a scythe shows him bent over in a posture that would leave him exhausted and crooked at the end of an hour. Today's cheap hardware-store scythe also pulls the body over in a stoop, a fatiguing position. One should be able to use a good tool all day, every day, without pain or abuse to the body. The great improvement in Roman Gaul of the long-handled scythe over the shorthandled scythe and sickle of early Roman times lay in the user's ability to stand erect and relaxed, and thus do more work.

My whole body twists around to the left, my left elbow bending further and leading the way around the back of my body. The stroke ends when the blade cuts through the line made by both feet (Figure 15, ending). At this point my left elbow is almost touching my backbone. My chest faces left. There is tension in every joint as the pendulum of the swing comes to its furthest point.

At its best, the stroke does not have to be stopped. It is initiated with just enough energy that the last of the grass is cut and thrown to the windrow as the momentum of the stroke is reduced to zero; the leftover energy is comfortably stored in the tendons to power the recovery back to the right.

As I swing back to the beginning of the next stroke, I move both feet forward a step. The heel and toe of the scythe are parallel to the ground and just above it at all times. The shape of the cut area is a crescent.

The grass tends to be thrown to the left by the scythe; therefore I proceed as if walking along the edge of the uncut grass. Many people confuse a scythe with a broom or a golf club. They face the field straight on rather than walk along it, and try to slash the grass. The heel of the blade always rises uselessly above the ground at the beginning of the stroke. Despite a vigorous swing, little is cut. The grass that is cut is thrown into the next swath. This approach is pictured in Figure 16.

For the right approach, I walk along the edge of the field, twisting my body, keeping the heel of the blade down to the ground from beginning to end (Figure 17). In golf or croquet there is only one point of impact on the ball. In mowing, there are a hundred points of impact and they must all be hit at the right height. Rudyard Kipling gave a nice metaphor for the sweeping arc of the mowing stroke: "The foresail scythed back and forth against the blue sky." The mower is the mast, the boom is the scythe. [13]

I practice on a lawn, dragging the blade in the stroke and in the return. In a cross-section of the blade from edge to rib, the area behind the edge is gently curved out

Figure 16

The hacking stance

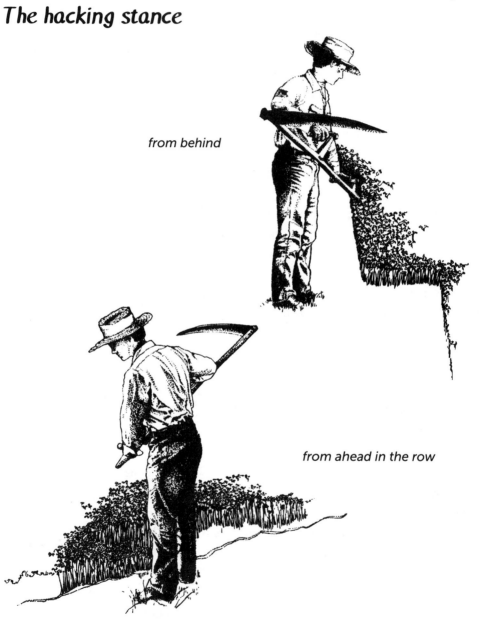

from behind

from ahead in the row

Figure 17

The rhythmic twisting of scything, blade always close to the ground

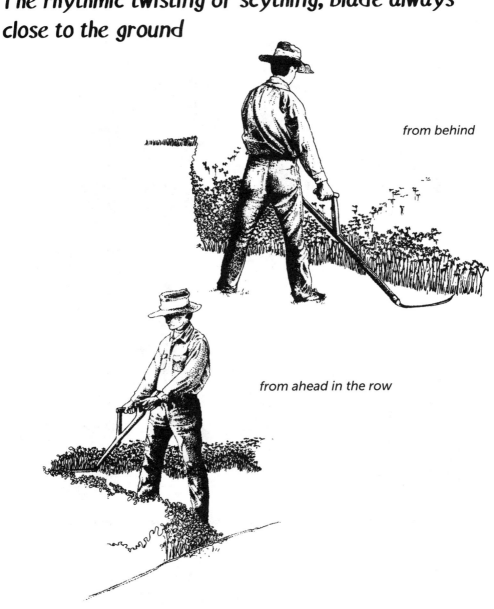

from behind

from ahead in the row

Figure 18

Reaching out to enlarge the swath

(Figure 2). In close mowing the convex blade slides along without digging as the stamped blade would. With this dragging exercise, the lawn gets mowed, and I can perfect keeping the blade parallel and close to the ground.

For a wider swath, my right and left hands reach out to the right at the beginning of the stroke and are further from my body in the swing. The mower painted in Figure 18 leans toward the grass without losing his balance and without stooping. An experienced mower in good grass can cut a swath seven feet wide or more with the 70-centimeter blade. An average swath six to seven feet wide was used as a unit of measure in hayfields in England. Inexperienced mowers would be wise to take smaller bites by walking a bit to the left, even walking in the cut grass of the previous row. It is no disgrace to be cutting a swath two feet wide. The importance of a clean cut over a wide cut is illustrated by the scoring systems of mowing contests. At the annual meeting of the Dairy Herd Improvement Association of Sullivan County, New Hampshire, 45 points are offered for neatness of the stubble, 30 points for width of swath, 15 points for time to cut one swath fifty feet long, and 10 points for making it to the end. In the Austrian national mowing competition, the time is recorded for cutting an area 10 meters square (1076 square feet), and time is subtracted for sloppiness of cutting. The width of the swath is not counted at all, which seems to me fairer to different sizes and shapes of people. The emphasis is on speed and the amount of cleanly cut grass, which is after all the aim of mowing. The same emphasis on a clean cut was made in the *Library of Useful Knowledge* in 1830: "The mowing should be so performed, that neither the strokes of the scythe nor the junction of the swaths can be discerned."

Most often it is not strength which permits a wider cut but rather the proper twisting movement to store energy in the tendons. The length and breadth of the swath also vary with the crop to be cut, its condition, and the evenness of the terrain. Poor grasses that are short and brown in August on hilly stony soil are obviously harder to mow than tall dewy greens in May on level ground.

SMALL ADVANCES

Although it is best for the beginner to stand erect without stooping over, he or she will have the experience of being pulled forward when the width of the swath is increased. When this position becomes a stoop, it should be abandoned; the knack of positioning yourself in mowing is to learn how to lean into the work without losing balance.

Even following all the rules of technique, there are still opportunities for variations. The snath and blade are manufactured to hold the cutting edge in the right relationship to the grass. The blade can, however, be manipulated to maximize the cutting angle. I cut, always, with the blade parallel to the ground; but this lets me try two different approaches to the grass. If I put my left hand against my belt and bring the right hand around through the stroke, I can make a cramped slice. The opposite approach is to pull the whole blade straight across the grass like a cleaver hacking meat. Neither extreme works. With the narrow slice, a mere two inches of grass is cut; with the broad hack, the grass tends to bend over. All mowing ideally combines the two approaches; in varying conditions I must alter the approach to have more slice or more hack. If the cutting is easy, I may reach out at the beginning of the stroke to increase the size of the swath, while sacrificing some of the slicing effect. If the cutting is difficult, I may wish to maximize the slicing effect and minimize the hacking effect by dosing the blade inward. Within the limits of the scythe ring, the blade itself can be shifted slightly in its seat to increase the slice or hack effects (Figure 5), but I have found that changing the way in which I use my arms, as just described, is adequate to adapt to most mowing conditions.

In some conditions I do not hold my feet stationary while cutting. I begin with my whole weight on the right foot; then, as my body begins to unwind, my weight moves onto the left foot. I move down the field in a slow walk.

These and other variations are appropriate in different conditions of cutting. Experiment. The grass will teach you.

The account given in *Anna Karenina* by Leo Tolstoy of the nobleman Levin joining the peasants for a day of mowing is a fine articulation of the ease and style inherent in mowing:

> *The old man, holding himself erect, moved in front, with his feet turned out, taking long, regular strides, and with a precise and regular action which seemed to cost him no more effort than swinging one's arms in walking, as though it were in play, he laid down the high, even row of grass. It was as though it were not he but the sharp scythe of itself swishing through the juicy grass . . .*
>
> *The longer Levin mowed, the oftener he felt the moments of unconsciousness in which it seemed not his hands that swung the scythe, but the scythe mowing of itself, a*

body full of life and consciousness of its own, and as though by magic, without think-
ing of it, the work turned out regular and well-finished of itself. These were the most
blissful moments.

It was only hard work when he had to break off the motion, which had become
unconscious, and to think; when he had to mow round a hillock or a tuft of sorrel. The
old man did this easily. When a hillock came he changed his action, and at one time
with the heel, and at another with the tip of his scythe, clipped the hillock round both
sides with short strokes. And while he did this he kept looking about and watching
what came into his view: at one moment he picked a wild berry and ate it or offered
it to Levin, then he flung away a twig with the blade of the scythe, then he looked at
a quail's nest, from which the bird flew just under the scythe, or caught a snake that
crossed his path, and lifting it on the scythe as though on a fork showed it to Levin and
threw it away.

For both Levin and the young peasant behind him, such changes of position were
difficult. Both of them, repeating over and over again the same strained movement,
were in a perfect frenzy of toil, and were incapable of shifting their position and at the
same time watching what was before them.

READING THE GRASS

The grass will teach you how to mow better than words on a page or proddings by a
teacher in the field. But you must pause to observe the grass if you are to learn its lesson.
Clearly speaking from his own experience, Tolstoy writes of Levin:

His pleasure was only disturbed by his row not being well cut. "I will swing less with
my arm and more with my whole body," he thought, comparing Tit's row, which looked
as if it had been cut with a line, with his own unevenly and irregularly lying grass.

Look back over what you have cut. Do the cut ends of the grass slope up to the right
into the uncut grass? The blade is too high at the beginning of the stroke. Keep it down.
Stop occasionally at the beginning of a stroke and look over at the blade to check its height.

Are there tufts of grass between the swaths? The blade is too high at the end of the
stroke, and you are probably trying for too wide a swath. You may also be throwing the

cuttings into the standing grass. Step to left, take smaller bites of grass, and finish the stroke all the way to the left.

Is there uncut grass in the middle of the swath? Sharpen your blade, and try more slicing.

Looking back over several strokes, does the swath have the appearance of rolling waves? The blade is probably not being held parallel to the ground, and the point is cutting at a different height from the heel.

And so on. If I cannot understand why the swath is uneven and irregular, I try to reproduce the badly cut grass, with exaggeration. Then I can discover the source of the trouble within my stroke, and correct my technique.

The grass will teach you, but it is easier to learn on lush green hay or weeds than on wiregrass and hardback. Young clover teaches an unstrained style as the blade whispers through the fragrant stems. The more difficult cutting of briars and tough weeds can then be approached with more effort but in a good style.

Improvement in mowing technique also requires reading the body with as much attention to detail as any athlete needs to apply.

Do your biceps ache from the work, or your back? Let your right arm hang straight. Twist your hips more, using the power in your ankles and knees.

Is your neck sore? Do not hunch up your shoulders, or stoop over, or work too hard on a wide swath.

Does your stomach hurt? Don't stop breathing. A common complaint is a soreness in the stomach because of the unaccustomed twisting. You are probably facing the field as in golf and should discipline yourself to keep the right and left feet perpendicular to the line of the cut, or even advance the right foot beyond the left.

Keeping the blade under the control of the body's twisting rhythm, rather than violently thrashing the grass, prevents exhaustion, and prolongs the mower's working day and his working years. When all the parts are working together, you will agree with the experience of Sir Stephen Tallents, who mows by "guiding my blade serenely through" the grasses.

Watch other mowers, and have someone watch you to see how small changes affect the cutting of the grass. There is room for personal differences in style. In some people the energy bursts from the left hip-bone, in others from the left elbow, in others it spreads evenly throughout the body.

THE USE OF THE BODY

Levin to his urban brother after his day of mowing: "You can't imagine what an effectual remedy it is for every sort of foolishness. I want to enrich medicine with a new word: Arbeitskur." Arbeit is German for work, movement, energy; kur means cure.

—Leo Tolstoy, *Anna Karenina*, 1876

Awareness of the dynamics of one's own body can be employed to improve one's ability at mowing, and, indeed, vice versa.

Most helpful for this understanding is the knowledge provided in the teachings of F. Matthias Alexander.

Alexander was an actor and public lecturer at the turn of the century whose career was interrupted by progressively worse vocal problems. Unable to find a medical explanation or cure, he spent several years carefully observing himself with the aid of mirrors, while speaking and at rest. He discovered that just before speaking, and even when thinking about speaking, he had been quite unconsciously tilting his head backward while jutting his chin forward. Though a small movement, it had cramped his neck and cut off the column of energy in his spine and throat, stunting his voice and presentation. At the same time, Alexander found, he would lift his chest and narrow his back, further cramping his larynx. After months of observation, he found that preventing the tension and movement in his head and neck resulted in a return of voice as well as an improvement in his general state of health. Asthma, which had plagued him throughout his life, diminished gradually, until the attacks ceased altogether.

Suspecting that his discovery of the head-and-neck's role in voice production contained even greater benefits, Alexander proceeded in his investigation to a scrutiny of the connection of head and neck to the entire human organism, and to the *use* of the body, or more accurately the *use* of the entire self. Working from the premise of the primary importance of the head-neck relationship, he was able to teach others to become aware of habitual *use*, thus freeing them for a re-educated use of their bodies more appropriate to the natural functioning of the human organism.

Alexander found that when he *tried* to keep a dynamic balance between head and neck and torso, his efforts *were vain*. His body became rigidly fixed in a new position. He also found that without mirrors he could not reliably know what he was doing with his head and neck, and therefore did not know if he was carrying them "correctly." Since he couldn't do anything to *remedy* the poor use, he was left with *not doing* what was hurting him. Inhibition of wrong habitual patterns of use thus became the central feature of Alexander's approach. In the words of a disciple of Alexander's:

Before an unrecognized pattern of behavior can be brought to your awareness, you must either recognize the habit, for example through mirrors, or an Alexander teacher's observation, or by "thinking yourself to stop" just when you are about to engage in an activity. Just when you are about to move, think "NO!" And, gradually, through repeated inhibition, rather than repeated habit, you develop an increased sense of kinesthetic perception, enabling you to sense yourself when "setting yourself" physically to do something.[14]

In my case, I first became aware of my poor use of myself in mowing; for example, I found that seeing dry wiregrass ahead in the row caused my neck and shoulders to tense, and thus put off my scythe stroke. I learned to inhibit this constriction during mowing in the way just described. Once I had inhibited the habitual response, I had created a pause between the stimulus (perception of wiregrass) and the response (tensing up for the anticipated harder mowing). In that pause is the opportunity for re-education of my use of myself. In that pause, Alexander suggested a series of three directions, to be given to myself silently as orders:

1. Let the neck be free (do not increase the tension in the neck),
2. Let the crown of the head go forward and up (do not pull the head back or down),
3. Let the torso lengthen and widen out (do not shorten and narrow the back).

These instructions sound simple enough, but they are nearly impossible to enact each by itself. Taken together, the three instructions make up one unified sensation, which Alexander called "the true and primary movement in each and every act."

The term "primary movement" is a reminder that we are always moving in response to the accelerating force of gravity, and that the head-neck relationship is the controlling center of this response. Alexander also termed this the "primary control" to emphasize how important inhibition was to suppressing habitual wrong patterns and permitting the body to do what it knows best. I use the term "primary movement" without ignoring the sense found in "primary control."

Primary movement can be observed in any small child, as well as in top athletes. Its loss leads to abuse of the body, with attendant symptoms of impaired functioning. To aid the client in rediscovering primary movement, the practitioner of Alexander technique uses the lightest direction of his or her hands, most often in the region of the head and shoulders. One can relearn to stand in a state of equilibrium, without tension (without military "Attention!") and without slumping under the burden of one's own weight. The experience of the body's antigravity power used in the twisting exercise before mowing is an example of positive use of the primary movement in relationship to the earth's power.

The anthropologist Raymond Dart used the name "poise" for "primary movement." By poise he meant a never-fixed dynamic equilibrium that is in effect even when the body is at rest. He related poise to the balanced energies in the double spiral of voluntary muscles twisting in opposite directions through the entire length of the body.[15] The twin serpents coiled about each other in the insignia of the medical profession effectively illustrate this same principle.

At the beginning of the scythe stroke my body is twisted to the right from toe to crown; at the end to the left. In each stroke I pass through the centering point of relaxed poise; at this moment, the blade is slicing through the grasses, and I am tempted to compress and tighten the muscles of my neck and back. It is worth the extra time to learn to be at my freest and loosest at this moment, while directing (ordering) my "neck to be free," "head to go forward and up," and "back to lengthen and widen."

The scythe grips are held by the hands only and need not involve the arms and shoulders in tension. The "primary movement" of head freed from the spine, forward and up with the crown, leaves abundant energy for the twisting through the rest of the body. The tool can swing freely, powered equally by every joint, and thus freeing them all.

The intrinsic rhythm of tool and mower combined, like a twisting and untwisting clock wheel, has a power and creates a timeframe all its own. A British farmer speaks of this experience:

> *I come to feel an indolence in action, letting the scythe do it, and a force that is not me takes possession of us both and we swing, swing.*[16]

Again we consult Levin:

Another row, and yet another row, followed—long rows and short rows, with good grass and with poor grass. Levin lost all sense of time and could not have told whether it was late or early now. A change began to come over his work, which gave him immense satisfaction. In the midst of his toil there were moments during which he forgot what he was doing, and it all came easy to him, and at those same moments his row was almost as smooth and well cut as Tit's. But so soon as he recollected what he was doing, and began trying to do better, he was at once conscious of all the difficulty of his task, and the row was badly mown.

We see that Alexander's "primary movement" has been well expressed by others, and we can encounter another aspect of the best use of the body, how it can change our perception of time. When I am using my body well, I enter into this other timeframe and follow the rhythm of my primary movement in relationship to the scythe, to the grass, and to the earth.[17]

Experiences recorded in Michael Murphy's *Golf in the Kingdom* and Eugene Herrigel's *Zen in the Art of Archery* reinforce Alexander's lessons and articulate them in the spiritual realm. The Zen paradox is that if I give up the hurried intention of "getting the hay in," I will in fact cut better. From Alexander's point of view, the hurried mower has forgotten the best use of the body in all the rush. It is so tempting to orient to the goal and forget the means we employ to get there. Always end-gaining, we forget what is actually happening. My objective is to have the hay; my attention is on the process of cutting it: I set myself at the beginning of the swath and let the mowing commence. "God speed the mowing!" as a form of salutation among mowers, much like "Good morning to you," seems to me to be an invocation to promote primary control.

Inhibiting my habitual tensions and permitting myself to enter into the clock wheel movement with the scythe can, however, be done while I am unaware of my surroundings. Then I am so entranced with the inexorable rhythm of my own primary movement that I do not look back on my work to check the stroke, and variations in the rhythm, such as are demanded by molehills, stones, and the rare toad or partridge, easily destroy the heavenly balance. Entering into a poised rhythm *with* awareness of my environment permits me to perform as evenly and flexibly as Levin's old teacher.

The human rhythm of mowing, and the flexibility it permits, are central to the superiority of the scythe as a tool. The first mechanical mowing machines used one or several scythes whirling about in complete circles from a horizontal wheel directly powered by the wheels of the machine turning on the ground. The inventors were trying to do what the human mower was doing, only faster. In fact, these machines performed worse than humans—they were unable to accommodate to changes in terrain, type of grass, direction of lay in lodged grass, and obstacles. On any but the smoothest homogeneous ground, the blades often dug into the ground or bent the grass over without cutting it.[18]

RECONSIDERING THE TWISTING EXERCISE

The meditative repose in which he performs (the preparations for working) gives him that vital loosening and equability of all his powers, that collectedness and presence of mind, without which no right work can be done.
—EUGEN HERRIGEL, *ZEN IN THE ART OF ARCHERY*, 1953

I again stand in a relaxed manner with the scythe on the ground in front of me. I take a moment to relax. Then I let my ankles become looser, so that I begin to rotate from side to side. I loosen my knees, hips, and torso, letting my arms flop from one side to another. I am always aware of the relationship of my head to my neck and to the rest of my body.

To this point I have concentrated on my own primary control and on inhibition of my habitual ways of moving, so that my body can twist more freely. Now, without actually doing it, I think about holding the scythe in this movement. If I tense up, then I can take the time to relax that tension without the real distraction of the scythe in my hands.

I then pick up the scythe and continue the twisting experiment, feeling the weight and shape of the tool. Finally, I can approach the grass with the scythe. In this way I proceed carefully from my relationship to gravity, to my own primary control, to my relationship with the tool, to my relationship with the grass. The scythe draws my primary control out beyond my own body to relate to the earth.[19]

Mowing Hay

Grass is the forgiveness of nature—her constant benediction. ... should its har-vest fail for a single year, famine would depopulate the world.
—Senator John James Ingalls, "In Praise of Blue Grass," 1872

UNLIKE THE LEAVES OF OTHER PLANTS, THE LEAVES OF GRASS (FAMILY *GRAMINEAE*) grow not from their tips but from cells at the base of their sheaths. Therefore they can be cut, mown, and nibbled again and again and yet grow back. "Greenstuff" is harvested to feed animals and to mulch gardens and orchards. Modern farming equipment has made the harvest easier, so that in place of the teams of mowers there is the tractor operator, his mowing machine, tedding machine, baling machine, and trucks for transport. Smaller farmers, gardeners, and homeowners have difficulty with haying because they cannot afford this equipment, cannot rent it in the suburbs, and cannot hire it from the larger farmer, who needs it at the same season.

Hay is cut first in the spring when the leaves are green, juicy, and full of protein, and when the fiber content of the plant is low. The digestibility and worth of the hay is at its highest quite early—May 15 in Vermont and earlier further south. Every day after this date digestibility decreases by half of a percent, as the proportion of protein to fiber decreases. Later on there is more quantity of hay as the grasses mature, but it is actually less useful to the animals. (Hay which is harvested for the first time in August has tough brown stalks without much leafy green: "Won't make butter, but it'll keep 'em alive," as one dairy farmer said.)

The reason mechanized farmers often wait until June and July to cut hay is that the weather has not settled and the fields are too wet for heavy equipment. When mowing by hand, however, there is no danger of getting stuck in the wet spots. Furthermore,

Figure 19

Mowing in Massachusetts in 1822

grasses at the peak of nutritive value are more succulent and easier to mow than older grasses. An early first cut lets the grass grow back for equally fine second and third cuts later in the summer (even more further south). These second cuttings (*rowen* or *after-grass*) are very popular with the animals.

A scythe is also more versatile than a mechanical mower along the edges and in the corners of the field. In many New England hayfields the tree line has crept onto the field, leaving the stonewall boundary ten or twenty yards back into the young forest. Tractor-drawn mowers are not designed to cut to the very edge. The scythe is more maneuverable and precise, and can mow in corners and close along walls. For very large fields, it could be used for margin-trimming in conjunction with the machines.

One thing harvested along the field's edge is small one-year-old trees. Having observed that all grazing animals spend so much time eating the leaves from the trees at the edge of the pasture, the authors of *Bio-Dynamic Agriculture* suggest adding a "leaf hay" to the daily feed ration as an added resource. So the scythe's trimmings along the field's edge are a boon to the hay.[20]

Historians report a greater diversity of plants and flowers in centuries past in the meadows used for hay and pasture. "Biological" farmers strive for a similar diversity, in contrast to the one or two kinds of plant found in the hayfields of modern largescale agriculture. One popular seed mixture from England for permanent pasture or hayfield contains nineteen different plants:[21]

2 lb.	Danish Roskilde Cocksfoot or Orchard grass (*Dactylis glomerata*)
3 lb.	Aberystwyth S. 143 Cocksfoot (*Dactylis glomerata*)
3 lb.	Aberystwyth S. 26 Cocksfoot (*Dactylis glomerata*)
2 lb.	Danish Meadow Fescue (*Festuca pratensis*)
1 lb.	Alta American Tall Fescue (*Festuca elatior*)
½ lb.	Rough Stalked Meadow Grass (*Poa trivialis*)
¼ lb.	Smooth Stalked Meadow Grass (*Poa pratensis*)
4 lb.	R. V. P. Italian Ryegrass (*Lolium italicum*)
7 lb.	Reveille Tetraploid Perrenial Ryegrass (*Lolium perenne*)
½ lb.	Canadian Climax Timothy (*Phleum pratense*)
2 lb.	Grasslands Huia White Clover (*Trifolium repens*)
1½ lb.	Alsike Clover (*Trifolium hybridum*)
1 lb.	A!taswede Red Clover (*Trifolium pratense perenne*)
½ lb.	Kent Wild White Clover (*Trifolium repens silvetre*)
1 lb.	French Chicory (*Cichorium intybus*)
2 lb.	Burnet (*Poterium sanguisorba*)
1 lb.	Sheeps Parsley (*Petroselinum sativum*)
½ lb.	Ribgrass (*Plantago lanceolata*)
¼ lb.	Yarrow (*Achillea millefolium*)

33 lb. per acre

The last five of these hayfield plants are herbs, sown in small quantities to provide a different sort of nutritional resource for the animals.[22] This is in fact the reason for including such a variety of seeds: from year to year and from week to week in season, one or another plant will find the conditions in which it will thrive using the soil to the maximum benefit of the animals that feed on it. Herbal hay is not only for ruminants such as cows and goats. For raising poultry, an ounce of varied greenstuff per chicken per day is the secret of absence of disease and healthy laying.[23]

The scythe mower has a final reward. The cut ends of each type of grass have their own unique perfume to add to the enjoyment of diversity.

MAKING THE HAY

My long scythe whispered and left the hay to make.
—Robert Frost "Mowing," 1913

An acre was regarded as the equivalent of a day's mowing. However, some could mow two acres a day without arising too early. An English legend is a man who, on a bet, mowed over five acres in one day; he used two scythes and employed a helper to whet the scythe not being used.

These measurements are relative to the type of "acre" being used. Our current acre contains 43,560 square feet; but the old Scottish acre was 1.27 of our acres, the old Irish acre was 1.62 acres, the old Cheshire acre was 2.12 acres, and several measures from Eastern Europe measure 1.42 acres. Each one may originally have related to the area mowed in a day.

The serious mowers would begin at four in the morning, when they could hardly see to whet their blades. The early morning dew gave the grass greater weight and inertia; it was less likely to bend than to be cut. Ease of cutting is not the only reason to cut low: for hay or mulch, there is more quantity in each stalk nearer the ground. Thus the saying, "An inch at the bottom is worth two at the top."

In England an attachment was used to improve the delivery of the cut grass to the windrow. Called a "grass nail," it was a stiff piece of wire connecting a hole in the beard with a hole three inches up the snath just higher than the tang. Another attachment used in different parts of Europe was a small semicircular bow with a four-inch radius attached to the base of the snath and rising perpendicularly to the line of the blade (much like the large bow cradle in Figure 26).[24]

Hay has been dried in a variety of ways. In one tradition, boys and girls followed the mower, strewing the freshly cut grass alternately with each hand over the opposite shoulder to lay it as lightly as possible. They came back with a hayfork to turn it again, exposing different surfaces to sun and wind. In Austria I observed the use of saplings with several branch stubs left. These eight-foot poles, called *heinzen*, were stuck in the ground, and the hay was draped around them using the branch stubs as hooks. Also in Europe, footcocks or small haystacks are made, sometimes on a tripod and platform to let air into the center. Three poles lashed at the top keep the hay up off the ground for better drying; such a tripod is shown in the background of Figure 20. The grass on the surface is combed with a rake so that rain is deflected from the bulk of the grass while it slowly dries. On small farms in the mountainous areas of Scandinavia, horizontal wires are rigged between support poles and the grass is hung up as on a clothesline.

Rain is not only a nuisance because it prolongs drying time; it can actually leach away nutrients from the surface of the plants. "He shall come down like the rain upon the mown grass'" (Psalms, 72:6). If rain threatens, it is best to gather up the windrows of drying hay into a large stack, and cover the stack. After the ground has dried, spread the hay out again. Or bring the hay under cover in a ventilated area and let it dry more slowly there before stacking it in storage.

The purpose of drying the grass is to preserve it. If it is too wet bacteria will rot and ruin it. The rotting also builds up a great deal of heat, and through history has caused the burning of many a barn. However, when they are too dry the very leafy crops such as alfalfa (lucerne), the clovers, and herbs will pulverize in handling and be left in the field or on the way to the place of storage.

When old-timers smelled the hay's deliciously heavy perfume in the field, they knew it was dry enough, and they snatched it up so it would lose no more of its essences. If left too long in the sun, the hay becomes light yellow, sometimes bleached almost white. The best hay is dry enough to rustle when picked up, but is blue green. The optimum range of moisture is 16–24 percent, but there is no way to know what the percentage of moisture in hay is without an expensive testing device. Finding the point where it is as dry as possible without being brittle is learned with a few trials. If the hay has 39–40 percent

Figure 20

Making hay

moisture, there is a fire hazard; a temperature over 180 degrees Fahrenheit is critical, and the hay should be spread apart or removed from the barn.

Horne's *Complete Grazier* (1830) gives a hay-making schedule designed for areas of unsettled weather. On the first day, the grass must be mown before nine in the morning. Then it is tedded or spread, every lump shaken out to expose the maximum surface area to sun and wind. Then it is turned again. In the afternoon it is raked into windrows and lastly into grass-cocks. The cocks, or little piles of grass, are combed on the outside to lead any rain away from the center. On the second day, the grass-cocks are pulled apart into staddles, then turned twice in the day, alternating with the tending of the grass newly mown that day. Then the first day's grass is again raked into cocks a little larger than before. On the third day, the grass is raked out into staddles and turned; if not dry enough to carry to the barn, it is raked into great-cocks until the next day. And so forth. Horne is accurate when he estimates that the work of twenty haymakers is needed for that of four mowers in areas where the drying is slow. Clarence Danhof estimated that in ideal conditions the work of tedding and raking took a half day for every day of mowing. Francis Alexander's painting of haymaking in Southbridge, Massachusetts, in 1822 (Figure 19) shows four mowers, three rakers, one cart loader, and one person carrying refreshment to the field.[25]

This working and reworking of the same piece of ground can lead to the sort of intimacy which Andrew Marvell described in 1681:

> I am the Mower *Damon*, known
> Through all the Meadows I have mown.
> On me the Morn her dew distills
> Before her darling Daffodills
> And, if at Noon my toil me heat,
> The Sun himself licks off my Sweat.
> While, going home, the Ev'ning sweet
> In cowslip-water bathes my feet.[26]

Going over the same piece of land again and again, learning where the obstructions are hidden, where the grass grows well and where poorly, is the original meaning of being the best in your field.

TOOLS FOR MAKING HAY

We shook the hay on our forks, whisking and whirling it shred from shred, scooping up more from the ground while that which was last on our forks was still falling, so that a continuous windy curtain of hay billowed before us.
— ADRIAN BELL, *SUNRISE TO SUNSET*, 1944

Come, lads, let us be merry and all of one mind.
And then we will expose the hay unto the Sun and Wynd.

—"The Months of the Year"
Nineteenth-century English song

On a small scale, the handling of hay requires a hayfork, hayrake, and some sort of conveyance (Figures 20 and 21). Hayforks were once forked sticks made from saplings which had grown into the right shape. Now forks are commonly available with three metal tines, which I find very useful in the barn or hay shed for pitching piles of hay from floor to loft and back again. In the field, however, I find a wooden hayfork, fashioned by hand along the grain of a single piece of hardwood, to be a more dependable tool. I can slide the blunter wooden tines along the ground and underneath the grasses I am trying to gather. The wooden fork is also wider than the metal one. Not everybody agrees; an elderly neighbor hayed as a boy one whole summer to receive twenty-five cents and a coveted two-pronged metal hayfork as his entire pay.

For tedding, the fork flicks the hay this way and that to fluff it up and fill it with air. For loading, the fork draws together the withered leaves and compresses the air out of them. To pick up stray wisps of grass, the fork is best used, not as a rake, but in a stirring motion to tangle the pieces of grass together.

Bull, buck, or drag rakes, occasionally four to five feet in width, were pulled along the ground to clean up the extra grass.

Rakes varied in their overall width, in the length of the wooden teeth, in the length of the handle, and the direction in which the teeth pointed depending on the terrain and type of grass. One type of rake has teeth pointing both up and down from the rake head, which attaches to the handle at a forty-five degree angle; by turning the rake over you can push the grass to one side or another into windrows rather than into a heap at your feet.[27] Another type is attached to the belt and dragged in that way. A long-handled hayrake 2½-feet wide, with fourteen teeth and a well-braced connection between head and handle, is still available, and is adequate for the job of cleaning the field. It is also easier to transport than some of the bigger bull rakes.

The best way to hold a hayrake can be observed in Brueghel's painting "Haymaking," as well as on ancient Egyptian scrolls: the thumb of the upper hand points up the pole while the thumb of the lower hand usually points down. In this way, the rake is not

Figure 21

Haying equipment

Hay Knife

48"

Modern Rake

Hay Pins

30"

one's height plus a fist

Forks

20"

angle sometimes
acute to 45°

30"

4' RAKE
(after Brueghel)

sapling

metal
3 tine

riven
3 tine

Figure 21

Haying equipment (continued)

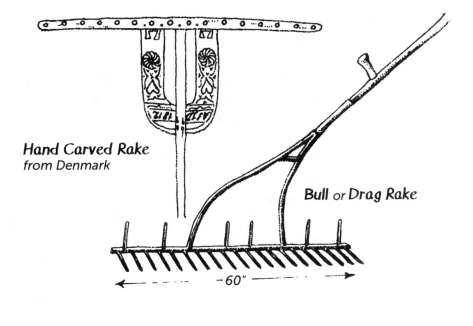

Hand Carved Rake
from Denmark

Bull *or* Drag Rake

← - - - - - ~60" - - - - - →

drawn across the ground where it might get tangled in the grass and break a tooth. Like a comb, the rake draws the cut grass *up and out* of the stubble.

If a tooth is broken, the broken stub must be drilled out and a new dowel of hardwood set it its place. Stronger replacement teeth can be made by hand, cleaving along the grain. In rake teeth or wooden pitchforks, hand cleaving exposes less end-grain to the weather than does a saw and makes the tool much more durable.[28]

Light vehicles for moving hay were common when power was provided by the draft animal. Modern motorized equipment is often not available, too heavy for a moist field in the spring, or unable to get to a certain place. Most garden and farm carts available in the United States have an unprotected space between the wheels and the body of the cart; when the grass falls into this area, it fouls the wheels. One sturdy cart design has detachable flare boards (also traditionally called "cart ladders"), which protect the vulnerable area between body and wheel and functionally increase the volume of the cart threefold.[29]

Since volume, and not weight, is the problem in transporting hay, the load can be increased even further with the use of hay-pins, long thin metal rods which, like hair-pins, were stuck down into the hay to hold the top of the load fixed to the grass beneath it. These pins must have eyes with strings attached for pulling them out of the load. Have you ever wondered how the missing needle in a haystack got there in the first place? These are the needles, and they must be retrieved for the next load of hay.

When loading a larger cart or truck, set each forkful of hay in a circle at the edge of the hay bed. Make the circle smaller as the load gets bigger, so the bulk of the hay tilts toward the center. Often there is a helper on the growing pile who sets the forkfuls in the best place for a large and stable load. At the barn the same person should take the forkfuls apart in the very same order in which the load was made.

When loose hay is forked into the hayloft or haymow, it naturally settles over time; some farmers encourage this by walking on it to exclude the air and thus better preserve the dehydrated greens. When the hay is compacted, it can only be taken out the way it was put in, a forkful at a time from the top. Long two-handled hay knives can be used to make vertical cuts in the hay to get to it from the side or to make exact divisions. In 1939, eighty-six percent of the hay cut in the United States was put up loose, so it is still within the memory of living farmers.

MOWING IN TEAMS

A scythe-sweep, and a scythe-sweep,
We mow the dale together.

—WILLIAM ALLINGHAM, *THE MOWERS*, 1865

The mowers arrange themselves at intervals along the edge of the field. The best among them begins the cutting in a relaxed rhythm which will quicken as the day wears on. Each in turn follows a few strokes later, mowing one swath to the right of his fellow and leaving the grass to the left in even rows. The team progresses around the field in a

Figure 22

The progress of mowing

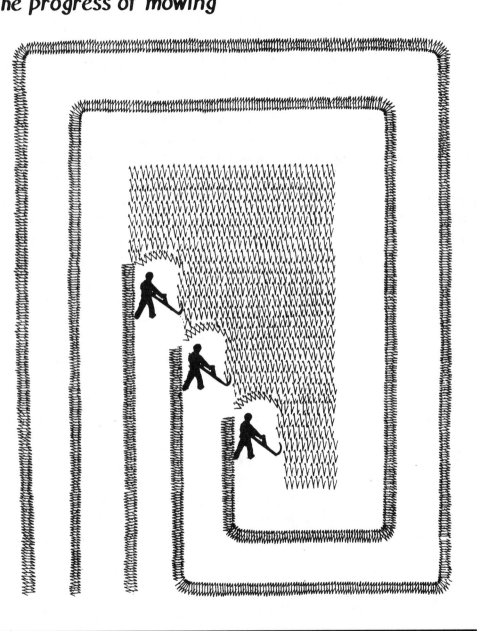

clockwise direction, each lap shorter than the last (Figure 22). At the very end, the outside edge of the field is carefully trimmed in a counterclockwise direction, throwing the grass into the field. Teams of three and four were a common sight in Britain at the turn of the century, and still can be seen in the high hills of central and eastern Europe. Levin in *Anna Karenina* mowed in a team of forty-two. What ground they must have covered in a day! In some areas farmers preferred to mow in teams with fellow farmers, moving from one neighboring farm to another.[30]

Some accounts speak of putting the best mower first to set a moderate but constant pace; the other mowers are compelled to keep up. Other accounts show the best mower put at the end of the row; the tickling sensation in the heels of the mowers in front of him keeps them going through the moments when they feel like pausing for a little conversation and rest. Even though some written accounts refer to the times when the mower behind yells, "Watch out, or I'll cut your legs off!", on the whole these teammates were well coordinated and quite used to working together. Making mowing a competitive match is unfair to beginners and usually ruins both the grass and the mower's primary movement. There should be more space and more leeway among amateur teammates.

Teammates also share their common rhythm through song. Though often heard among mowers, none of these songs has been recorded, probably because they were simple ditties, half nonsense. One hint of what the songs may have sounded like is the mower's lament at the food his master was providing, muttered in time with the stroke: "Bread and tea, no good for me." If the diet was changed, the chant would be, "Ham and eggs, look out [for] your legs." The main function of these chants and charms and songs was to coordinate the breath and scythe stroke, and to "deceive the tedious time, And steal unfelt the sultry hours away."[31]

With practice, a mower can learn to mow in relationship to the body, the tool, the grass, the earth, *and* the team of mowers. Many relationships of which to be aware! This expertise cannot be hurried but must be grown into. As the separate mowing rhythms of the different mowers synchronize through experience, one can discover an intimacy of shared work that is as thrilling as the discovery of good use of the body in relationship to the grass.

HAYTIME DRINKS

Oh, let the cider flow
In plowing and in sowing—
The healthiest drink I know,
In reaping and in mowing.

—D<small>EVONSHIRE WORK SONG</small>

Mowing would begin very early in the morning and continue until about eleven. Then the mowers would break for beer, cheese, and bread, followed by a nap, work to resume two hours later.[32] They might work until dark. Refreshment was the reward and the fuel, and Irishmen hired out for the haying season expected to be paid partly in beer—a forty-three gallon barrel of it for each mower. Unless they had to, mowers would avoid drinking before the midday break. But at that break!

> *There is the moisture he craves and a cask with a tap already inserted. Beer! His wife meets him and hands an uncorked jar. He nods briefly, draws his hand across sticky, clammy lips, and lifts the familiar vessel. . . . The gods must have had attractive temporal joys to offer mortals if they had better things to give them the all-absorbing ecstasy of that draught . . .*[33]

Also drunk profusely at the midday break was hard cider, which had a similar alcoholic content to beer and brought on a good sleep before the afternoon mowing. Hay is usually needed for wintertime feeding in areas where apples grow well. Good cider is made from many different kinds of apples, including a small proportion of crab apples, and is aged in oak barrels to improve its flavor. It is a wonderfully refreshing drink.

Still another popular pick-up tonic for haytime was switchel:

1 cup maple syrup, honey, or brown sugar
1 cup apple cider vinegar
½ cup light molasses
1 tablespoon ground ginger
1 quart cold water

Combine and stir well. Makes about 6 seven-ounce glasses. Hertzberg, Vaughan, and Greene, in *Putting Food By*, recommend switchel for modern athletes as a "good energy-restorer without promoting 'cotton mouth.'"[34] These drinks provided one of the rewards for the very long days of labor put in at haying time. The modern mower can enjoy them, too.

Clean Culture

Springing herbes reapt up with brasen sithes.
—Surrey's Aeneid, 1547

. . . every farmer mowed every bit of roadside, all around his land. Every bit of
grass was cut, around stonewalls, trees, buildings . . .
—Floyd Fuller of Randolph Center, Vermont[35]

WEEDS

IF I CONSCIENTIOUSLY CUT THE WEEDS AROUND THE GARDEN, ALONG THE ROAD-
side, and in every marginal area before they go to seed, I will slowly improve my home-
site. A favorite nesting site of garden pests is removed; over the years, the legions of
weeds will retreat. Every European farm, large and small, which I have visited had at
least one scythe; on several highly mechanized farms, the scythe's only use was extensive
trimming. Such obsessiveness is not a quaint trait of the old country. Every minute spent
with the scythe in this manner is repaid double in minutes saved weeding.

When cutting weeds, the scythe must be controlled more carefully than is necessary
in mowing. The strokes must be shorter and very accurate to shave around stones, trees,
fence posts, a bird's nest, a flower desired to be left. I am slowly learning to use the point
of the scythe to make the finest distinctions between blade of grass and the trunk of
apple tree.

In "The Tuft of Flowers," Robert Frost describes his discovery of "a tall tuft of flowers
beside a brook,"

A leaping tongue of bloom the scythe had spared
Beside a reedy brook the scythe had bared.

The mower in the dew had loved them thus,
By leaving them to flourish, not for us,

Nor yet to draw one thought of ours to him,
But from sheer morning gladness at the brim.[36]

An opportunity for discrimination and observation is given with the scythe that is not available from its mechanical substitutes. Many weeds have great virtues, and when kept under control with the scythe can be more deeply appreciated. Dandelions make a fine wine, milkweed pods make excellent fritters, lamb's quarters make a very nutritious early salad green (which gives this plant another name, "wild spinach"), goldenrod makes a simple dye for children's school projects, and so forth.[37]

When cutting down weeds in different places around the home or farm, I ask the question: Why is this particular species growing here? Ehrenrend Pfeiffer wrote a short pamphlet to help answer this question, entitled *Weeds and What They Tell*. The closeness to the earth which the scythe permits, compared with a tractor, makes identification of weeds possible. The information is available for interpretation, and may help in decisions to improve the land.

PASTURE

In pastures the nutritive and sweet-tasting plants are eaten first, then the less desirable. Finally there remain the plants that the animals won't eat. These mature, go to seed, and eventually take over. The animals eat themselves out of a food supply. Good pasture management involves not only attention to the mineral balance of the soil and periodic rests of the land rotation, but also getting the worst plants down before they go to seed. A scythe can do this quickly, even in large fields, since the mower has more maneuverability than a tractor when going from one dump to another.

LAWN

The emphasis in a lawn is on what is left after the cut grasses are taken away; therefore, lawn-tending is a variation of the clean culture approach. Until the invention of the

Figure 23

Mowing the grass

hand-pushed rotary lawn mower in 1831, all lawns were kept trimmed by nibbling animals or by gardeners with long-bladed scythes (Figure 23). From reports of travelers, the scythe is still used in Greece and Turkey, and perhaps elsewhere as well, for cutting lawns. To cut my lawn of eight thousand square feet, the 70-centimeter scythe requires an hour and a quarter; a hand-pushed and a gasoline-powered mower both average about forty minutes. Considering the cost of new equipment and of maintenance, and granting myself a reasonable wage for my labor, the scythe is economically competitive. It is also free of noise and exhaust, and free of the dangers of gasoline-powered mowers (the Consumer Product Safety Commission estimates that each year 77,000 people are cut or lose a finger or toe to lawnmowers). The scythe would be still more economical on a smaller scale, but not for the modern golf course.

For the border of the lawn, the new gasoline-powered scythe substitutes feature a short nylon cord that whirls around at the end of a three-foot handle held in the hands. The whirling nylon beats the grass blades apart, resulting in bruised rather than cleanly cut leaves. The nylon also gets beaten apart, though more slowly, and is added to the litter at the lawn's edge. I have many times observed men using these tools very effectively, as fast as they could learn to wield a scythe. I have also seen them many times attacking small tufts of weeds, passing the tool back and forth obstinately over stems that lie too flat to be bludgeoned while the operator is treated to the high whine of the machine and the billows of oily blue-gray smoke. With the scythe, I would have leaned over to pull the difficult weeds by hand, or I would have returned with an edging hoe whose blade extends straight from the handle to cut through the sod along the line between the lawn and something else.

One situation in which the scythe and rake definitely have the edge over conventional lawn mowers is a lawn that has been neglected and grown to a hayfield. However, an article by George Taloumis entitled "Cutting a Lawn Gone to Meadow" makes no mention of the scythe. Instead, for his eight-thousand foot backyard he hired a lawn service which used four gasoline-powered tools: self-propelled mowing machine large enough not to be clogged by the tall grass, a nylon cord trimmer, an air rake, and a truck to haul them. The job required nearly four and a half hours.

Mowing the lawn by hand is easiest when the grass is heavy with morning dew. I stand with my feet wide apart to give a wide swath, and drag the blade along the ground. To raise the height of the cut I raise my left hand during the stroke still letting the blade drag on the ground. If I follow the blade with my eye, I can see it moving through the grass in a way which is obscured in the hayfield; it looks like a fish speeding through water seen from above the surface.

The scythe leaves the lawn grasses in windrows which must be raked up, or more traditionally swept up with a besom of tree branches. In August these are the greenest and most succulent grasses on the farm, and I feed them to the animals as "green chop." At the end of the season, I am relieved of the chore of thatching the lawn—that is, raking it firmly to pull out the dead grass cuttings left by the power or hand-pushed lawnmowers. However, hillocks in the lawn are frustrating to the scythe, and I am drawn to the extra task of rolling the lawn in the spring to make it more even.

Harvesting Small Grains

Sowing in the morning, sowing seeds of kindness,
Sowing in the noontide and the dewy eve;
Waiting for the harvest, and the time of reaping,
We shall come rejoicing, bringing in the sheaves.
—GEORGE MINOR AND KNOWLES SHAW, NINETEENTH-CENTURY HYMNS[38]

FROM STUDIES OF THE MODERN AMERICAN DIET, IT SEEMS MANY PEOPLE HAVE forgotten an axiom known to agricultural man until just one generation ago: Bread is the staff of life. Bread in all its forms—from pancakes to hardtack to croissants—is grain made edible. While many of our children know grains only in the form of sugared break- fast cereals and hamburger buns, farmers not long ago were familiar with a diversity of grains which they used for specific purposes. They knew what was best to eat for them- selves and their animals under certain workloads and in certain weather conditions. And they knew what was best for brewing.

The modern gardener or small farmer can easily raise a small quantity of grains for his or her own table. It is worth a try. With an expenditure of a couple of hours on a good patch of soil 38 feet square, you can grow a bushel of wheat, which will make 1,792 four-inch pancakes (at the rate of 30 bushels/acre). Other sections for rye, oats, barley, buckwheat, and so on add different tastes and different experiences. By becoming famil- iar with the different sorts of nutrition these grains offer the body, you can make a link with the heritage of agricultural use of the land leading back into Neolithic times, ten thousand years ago.[39] Mahatma Gandhi added a moral dimension to growing one's own: he insisted that each person should contribute a portion of his or her labor to his or her own sustenance; he called it "bread labor."

Ever since grains (which are also of the *Gramineae* or grass family) have been harvested, a scythe-like tool has been used, that is, a tool with a cutting edge of stone or metal attached at a right angle to the handle. Europeans, who call the local grain of any area "corn," discovered a very different sort of corn cultivated in the American continent. They called this "Indian corn" or "maize." It is the only grain crop that cannot be harvested with a scythe, although the traditional "[Indian] corn knife" is a very sharp twelve-inch blade hafted at a right angle to a short wooden handle.

This section first gives a brief history of harvesting tools for small grains, then a description of how small grains are produced, and finally a brief illustration of some old and new harvest customs.[40]

THE SCYTHE AND ITS PREDECESSORS

At first, grains were harvested by pulling the whole plant up by the roots, or the kernels were knocked off into a basket. Early cutting tools were animal jawbones with sharp chips of flint stuck in where the teeth had been, or crooks of wood, also with flint chips. These designs all used the idea of a hook—the tool was passed in behind the standing grasses and jerked towards the body. Early bronze sickles were also hafted at a right angle to a handle of wood or bone. The blade was eventually made to curve back to the right, and then was continued in a long curve to the point at the left. Thus the weight of the tool was evenly balanced and the wrist was less strained after wielding the sickle accurately hour upon hour.

A short-handled scythe held with one hand permitted the reaper to work while standing erect. When using a sickle, the reaper's left hand grabbed bunches of straw to be cut. With the short-handled scythe, the left hand was extended with a short wooden crook to draw the cut grain out of the field as it fell to the scythe's stroke. An easily accessible illustration of this method can be found in the famous fifteenth-century *Tres Riches Heures* of Jean, Duke of Berry.

The *Hainault*, or Flemish, type of short-handled scythe had a blade over two feet long, and a snath fourteen inches long with a delicately carved hand-grip at right angles to the snath. The wooden crook, or *mathook*, was also carefully shaped to fit the hand. Stephens (*The Book of the Farm*) estimated that this scythe saved 25 percent of the work compared to the sickle. However, the Flemish scythe was much more dangerous to the

Figure 24

Ancient and modern harvesting implements, drawn to scale

a b c d e f g h

Figure 25

Flemish scythe and reaping hook

operator than the long-handled scythe held in both hands. John Gerard, in his seventeenth-century *Herball*, recorded an accident wherein the mower

> *made a wound to the bones, and withall very large and wide, and also with great effusion of blood; the poore man crept upon this herbe (Clownes Wound wort or All-heale), which he bruised with his hands, and tied a great quantity of it unto the wound with a piece of his shirt, which presently stanched the bleeding, and ceased the paine, insomuch that the poore man presently went to his dayes work againe, and so did from day to day, without resting one day until he was perfectly whole, which was accomplished in a few dayes.*

The Flemish scythe was widely used in the Dutch settlements in the Hudson, Mohawk, and Schoharie valleys of New York.[41]

It was in the undermanned Roman estates in Gaul, where greater efficiency was required in the harvest season, that the long-handled scythe was developed, originally longer than the scythe we know today, and without grips. Scythes with one grip first appear in illustrations from the twelfth century; this design enabled the mower to swing the full stretch of his arms, and thereby to increase the width of the swath. Later two grips appeared in most localities, further increasing the tool's efficiency.

In an analysis of probate inventories and wills, Georges Duby found that most medieval European farmers had a single scythe, often as their only metal implement. In the fourteenth century, Duby reports that a scythe in Haute Provence was worth a fortune. A presentation of probate records by John Roper shows that scythes were also valuable in sixteenth-century England. By comparing the appraised values for scythes and other items in the probate inventories, such as horses, swine, and heifers, and comparing with modern prices for the same farm animals, we can guess that a scythe blade at one shilling in sixteenth-century England was worth about $87 in 1981 United States currency. The expense was one reason why it was not until the eighteenth century that the long-handled scythe became the universal reaping tool in the Western world.

Despite advances in the efficiency of scythe design, when workers were unskilled and slow, or when mature grain had blown over in strong wind and heavy rain, the farmer preferred to use sickles since it was more difficult to damage the crop with sickles. In some areas, the tradition persisted to harvest wheat and barley with the sickle, oats and rye with the scythe.

CRADLE SCYTHES

They use a greater Sythe with a long Snath,
and fenced with a crooked frame of stickes,
wherewith with both hands they cut downe the Corne,
and lay it in Swathes.

—Conrad Heresbach, *Foure Bookes of Husbandrie*, 1577

The most elaborate and most beautiful invention for laying out grain more neatly to the side in scything is the grain cradle. According to thorough research by Jared Van Wagenen it was invented in New England in 1776.[42] It was really a completely different tool, quite like a scythe in principle, but with a rack of three to five wooden tines curved to follow the shape of the blade (Figure 26). At the end of the stroke, the straws would be bunched together and supported by the tines. The cradler would then tilt the whole thing to the left and let the cut grain slide out in a neat bundle. In the middle of the nineteenth century there were a million cradle scythes being used in the northeast United States alone; they have been used in the small fields of hilly areas in Appalachia and the Ozarks until quite recently.

Not all authorities agree with Van Wagenen that this cradle scythe was an American invention. Axel Steensberg claims that the cradle was developed in Scandinavia while other European authors claim different points of origin. Steensberg reported a reference in the psalter of Margrethe Skulesdatter in the beginning of the thirteenth century. He carefully analyzed the Danish probate records from the seventeenth and eighteenth centuries, and found that by 1767 cradle scythes accounted for half of all reaping and mowing implements. What Steensberg means by cradle scythe, however, is certainly a different design from that of the American cradle: a little rake of two or three teeth attached to the scythe ring, rising all of eight inches from the blade, the teeth extending six to fifteen inches along the blade.[43]

In the American cradle scythe, the craftsmen were able to raise the rack fully twenty-four inches from the blade, with the tines curved to follow the exact contour of the blade for thirty-six inches. The invention, therefore, is really a dramatic extension of an existing design. Bigger snaths and bigger cradles for a bigger country.

The American cradle scythe came in different shapes, most commonly a straight or "turkey wing" shape, and a very bent or "grapevine" shape. With the grapevine, the cradler must be quite stooped over to obtain the best angle of cut. In some localities the tradition is to balance the cradle with the right arm and left leg while the bundle of cut grain is scooped off the rack with the left hand and laid on the ground. This balancing act has always seemed to me to be a terrific amount of work, and hardly a good use of the body. With the turkey wing model use of the body is better; the cradler stands more erect, and at the end of the stroke the cradle is tilted to the left and jerked to dump the bundle of grain onto the ground.

Directions for constructing a grain cradle attachment to an "American" scythe are given in an article by Richard Weinsteiger. There are two problems with this cradle. First, it is quite difficult to construct properly—bending the wooden tines and fitting the pieces together required experienced hands, and fine cradles were marked with their makers' names (for example, "Christian Bouck" of Mineral Springs, New York). Secondly, the cradle attachment makes the "American" scythe even heavier and more difficult to work with.

In Europe cradle-like attachments were developed for the grass scythe to make it more efficient in harvesting grain. These attachments are usually a wooden or wire bow begin- ning at the blade end of the snath and attaching again to the snath just below the lower grip. Thus the cut grain stalks were not "cradled" or supported along the entire length of the blade, but were caught toward the end by this simple framework. The attachment was light, it could be detached when not needed, and the reaper's posture could remain as erect as it was in mowing. It seems that the time lost in somewhat less efficient piling of the bundles of grain would be more than gained back in the more efficient use of the reaper.

Today's European-style large bow cradle, shown in Figure 26, is mailed as a straight hardwood splint with two metal fittings on each end. One fitting is slipped between the scythe ring and blade and is held there by the two set screws in the scythe ring. The other fitting is bolted through the snath just below the lowest position of the lower grip so that it points slightly away from the blade. The reaper makes a framework of strings between snath and bow to catch the grain as in a cradle and to extend the strength of the snath to the relatively flimsy bow. The shape is adjusted to match that of Figure 26, and is further adjusted to accommodate the conditions of the reaping. With the thinner gauge of the two types of bow, a support string may be needed to come forward to loop around the tip of the blade. The heavier weight bow holds its own shape, but it must be molded

Figure 26

Cradle scythes *(not drawn to scale)*

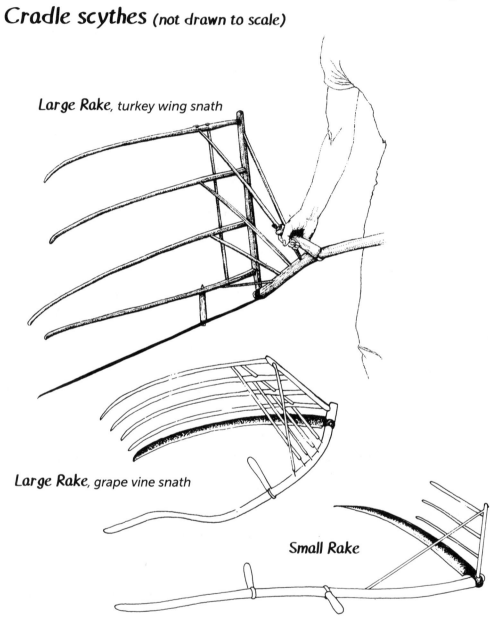

Large Rake, *turkey wing snath*

Large Rake, *grape vine snath*

Small Rake

Figure 26

Cradle scythes (continued)

Large Bow

F.A.O. Cradle

into that shape by steaming the wood for fifteen minutes and bending to the right curve before attaching it to the snath. For either type, when the reaping is over, the strings can be untied from the snath and the cradle stored until the next season.

Many variations of these basic designs have been found and are still in use. In the 1920s, the Seymour Manufacturing Company in Seymour, Indiana, offered sixteen different patterns of the American large rake grain cradle. The bow form of cradle has been found in every size, as small as a semicircle with an eight-inch radius for short grains or tall grass. It is unclear when and where each of the different designs originated, and exactly what crop condition each variant design was trying to accommodate. Just after World War II, the Food and Agriculture Organization in Rome introduced to Third World countries a detachable cradle for a straight snath with Austrian-pattern blade, but the cradle was too complicated to manufacture and repair, and the emphasis in the "Green Revolution" soon came to be placed on bigger and better machinery.[44]

CUTTING GRAIN WITH OLD AND NEW TOOLS

In his fascinating study of modern and ancient harvesting implements, Axel Steensberg performed comparative time trials using a variety of sickles and scythes from all periods of agriculture. In August of 1938 and 1939 he cut mature barley and oats with each tool from an area 5 meters wide and 10 meters long (538 square feet). Steensberg actually counted the number of straws in each plot, and arrived at an average figure of 26,646! In Table 1 I have given brief descriptions of the tools he used and the time it took to harvest these matched plots. When Steensberg cut more than one plot, I have given the repeat times; this shows some of the variations that can occur, apparently randomly, under identical conditions.

All of the flint implements in Steensberg's tests were tools actually used over six thousand years ago, which had been dug up by archeologists. They ranged from 4½ to 7 inches in length. The Stenild flint sickle (11.6 centimeters or 4½ inches long) is famous because it was recovered at Stenild in Jutland with the handle intact, and so could be copied exactly (*c* in Figure 24). It is dated in the early Bronze Age, approximately six thousand years ago, since the haft showed signs of being worked by metal. The dates of the bronze implements range between 4000 BC and 1500 BC, and have edges between 6 and 8 inches long. The Roman short-handled scythe found at Uggerby at the northern tip of Denmark (*e* in Figure 24) is more recent, perhaps dating from the first century before Christ; its blade is 30.5 centimeters (12 inches) long, and its handle 61 centimeters (24 inches). The Norwegian short-handled scythe, of the age of the Vikings, is more recent still, with a blade about 18 inches long and a handle 31 inches long, held between 18 and 22 inches from the blade (*f* in Figure 24).

The hafts of the tools in Steensberg's tests ranged widely:

1. No haft, the blade being wrapped in cloth or hide to protect the hand (*a*).
2. A crescent-shaped piece of wood looking like a boomerang in which the cutting edge was embedded (*b*).
3. The "knob" type, where the bronze blade was set in the end of the handle so that a portion protruded to the other side, suggesting an early form of poll to balance the blade's weight (*d*).
4. The Roman and Viking hafts, in which haft and blade were joined much as in modern scythes (*e* and *f*).[45]

In August of 1979 and 1980, I performed time-trials of four different methods of harvesting oats and barley, and have given the times, based on a plot size of fifty square meters, in Table 1. I cannot be sure if there were fifty straws in each square foot, as in Steensberg's trials, since I did not count. My stand was, however, too thick with grain, and had clovers and herbs undersown, so the harvesting was probably more difficult than in Steensberg's case.

The general indications are clear enough: Improvements in design have improved performance. Some of the ancient tools are amazingly efficient for their crudity, although I do not know how the reaper would feel after a full day's work, since several of these implements required seven or eight strokes to cut through a handful of straws. Rating the continuous output of the human body at 1/20 horsepower, all the hand-tool performances produce more per unit horsepower than the machine, even in comparison with the modern 190-horsepower combine, which takes twenty-four feet of standing grain in at one end, and delivers the grain threshed, cleaned, and bagged at the other end at the rate of one bushel of wheat in twenty seconds.[46] It is important to consider, in this context, the time and energy required to maintain a hand-tool compared to the time required to maintain a tractor or combine.

Table 1. Minutes required to cut barley and oats in 50 square meters (538 square feet), using ancient and modern harvesting implements.

Steensberg's time trials:	MINUTES
One-piece flint sickle: unhafted, held in the hand (a in Figure 24)	95,76
Stenild flint sickle, single flake embedded in haft of birch at right angle (c)	76,101
Crescent-shaped flint serrated "saw": flake set in slightly bent piece of wood (b)	73, 59, 72
Crescent-shaped bronze serrated sickle: bronze edge set in crescent-shaped piece of wood	65, 69
Smooth-edged "knob" bronze sickle, two types: curved metal blade at right angle to haft (d)	63, 64
Modern balanced sickles, two types (g)	30, 31
Short-handled Roman scythe of iron, straight 24-inch handle (e)	30
Short-handled Norwegian scythe of Viking Age, straight 31-inch handle(f)	17

My time trials:	MINUTES
Modern serrated American sickle (g) (Grain laid neatly by handfuls for ease in binding)	23, 27, 38
Modern scythe with "Austrian" hammered blade (h) (Grain laid in slightly unaligned rows requiring more work in binding)	6, 6, 9
Modern scythe with "Austrian" blade, European-style cradle (Grain laid in evenly aligned for ease in binding)	6, 7
Modern cutter bar, powered and hydraulically lifted gasoline-fed tractor (30 hp) (Grain laid every which way requiring sorting and realigning in binding)	2, 3

I began two other plots with old American cradle scythes, a turkey wing and a grapevine. Both broke before the plot was finished, renewing my sense of caution with old wooden tools left to sit.

Others have measured the area covered by a cradle scythe. Thomas Jefferson's diaries record an average of three acres reaped a day for the cradle scythe. Stephens, in *The Book of the Farm*, reported 2.75 acres of wheat and 4.125 acres of barley and oats per reaper in a ten-hour day.[47] Translated to the units of barley and oats of the time-trials in Table 1, a rate of four acres in ten hours comes to an average of two minutes per fifty square meters! The actual time necessary to reap an experimental plot must await further time-trials.

In my own experience, the Austrian-style scythe without cradle is faster than the American cradle in cutting, the latter being heavier and more cumbersome (average weight, eight pounds); the cradle, however, does leave less work for the gatherers and binders. The European-style bow cradle also leaves the grain in neat piles, but it is much lighter (average weight, four pounds) than the large rake cradle, and to my surprise, was no slower than the scythe without cradle in cutting.

SOWING, REAPING, THRESHING, WINNOWING

Sowing in the sunshine, sowing in the shadows,
Fearing neither clouds nor winter's chilling breeze;
By and by the harvest, and the labor ended,
We shall come rejoicing, bringing in the sheaves.

—GEORGE MINOR AND KNOWLES SHAW

Small grains are best sown in an area which has been cleaned of weeds, preferably by growing root crops in the plot the year before. You must keep the weeds out of the root crops through cultivation; the grain will benefit from this clean culture. Excess nitrogen and lack of potassium causes *lodging* or falling over of the grain, making harvesting difficult. A cradle cannot be used on lodged grain, and lodged grain must be harvested soon with a scythe or sickle to keep the kernels from sprouting by being too close to the damp earth. Lodged grain can be raked and tedded like hay.

Grain can be sown in the fall or early spring, depending on the climate and on the farming schedule. If grain is sown too late in fall, or too early in the spring, the cold and damp soil will rot the seed before it germinates. Though higher yielding, winter wheat sown in August or September can be completely killed by a hard winter. The finest baking flour with the highest gluten content to promote rising is spring-sown hard red wheat, in varieties such as Neepawa, Olaf, Manitou, Justin, and Selkirk. Barley, oats, and buckwheat are also sown early in the spring because they are not winter-hardy. Rye can be sown nearly any time since it is very hardy and requires only a short season for growing; the saying "wheat for fat, rye for strength" may indicate more the types of people who live in short-season growing areas than the nutritional quality of the grain. The seed can be broadcast by hand from a sack, though a rotary hand-turned seeder will give more uniform results. Cross the field back and forth in one direction, then again in a direction perpendicular or diagonal to the first for best results; an acre requires an hour by this method. One hundred pounds per acre is an average seeding rate.

After sowing, the seed must be covered with soil by raking the area, to improve germination and to protect the seed from the birds. The hayrake serves here as a grading rake, and it is best to step in the raked earth while raking, because the feet press the covered seeds into the earth. For larger areas, there are machines which spread the seeds, bury them, and firm the earth in one pass.

It is often a good idea to sow clover or alfalfa at the same time. When the grain is harvested the slower legumes will grow into autumn and give the field a protective cover through winter and a hay crop in the spring.

On the efficiency of sowing by casting the seed in all directions, by hand or with a whirling machine, George Fussell cites the parable of the sower:

> *Behold a sower went forth to sow; and when he sowed, some seeds fell by the wayside, and the fowls came and devoured them up; some fell upon stony places, where they had not much earth; and forthwith they sprang up because they had no deepness of earth; and when the sun was up, they were scorched; and because they had no root they withered away. And some fell among thorns; and the thorns sprang up and choked them; but others fell into good ground and brought forth fruit, some a hundred fold, some sixty fold, some thirty fold.*

Fussell used this story as evidence for the superiority of seed drills, which plow a small furrow, drop seeds into it, cover them over, and firm the earth above.[48] If a seed drill is available, such as the old and dependable Planet Jr. row planter, the grain can be sown thickly in rows seven to twelve inches apart from each other.[49] The drill can be adjusted for seeding rate, depth of furrow, and distance between rows, all determined by experimentation within the limits of soil and climate. The space between the rows can be cultivated or hoed once or twice in the growing season to suppress competition from weeds. Yields in row-sown plots are often better than with broadcast seed, and there is less problem of weeds in the mature grain. Row-sowing is a popular method among organic farmers in Europe. There is definitely more work involved, however, and, if a patch of land is free of weeds to begin with, I use the broadcast method, and read the parable of the sower as a parable.

In 1911, F. H. King, former Chief of the Division of Soil Management of the United States Department of Agriculture, published his memoirs of a tour of the Orient, where he found wheat grown in hills or groups averaging forty-six stalks each. The hills were twenty-four inches apart in the row, the rows were in pairs sixteen inches apart, and each pair of rows was thirty inches from the neighboring pair. The spaces in between the hills were hoed conscientiously after every rain to eliminate weeds, to incorporate frequently applied composted human wastes, and to create a moisture-conserving fluffy dirt mulch. King observed a harvest in drought of 12 bushels per acre, but calculated that the fine harvest of 1901 measured 116 bushels per acre.[50]

When should grain be harvested? Farm books give the infuriating answer: When it's ready. Working with other farmers and raising grains myself, I was surprised to find that there is no better answer. When you press a fingernail into the kernel of grain you are about to harvest, it shouldn't squirt out its contents as in the "milk" stage. But when it is really hard it is also ready to fall off; then the grain head shatters when the plant is cut, and all the kernels fall irretrievably to the ground. Rub a seed head between both palms to determine how fragile it is; chew it to determine if the kernels are at the "hard dough" stage-ready harvest.

With either a scythe or a cradle scythe, the stroke for reaping is broader than the mowing stroke, with more hacking and less slicing. With the cradle, the left hand is as close to the ground through the stroke as is comfortable in order to keep the cradle structure upright. The right hand comes up at the end of the swing, and the left hand drops abruptly to dump the load with the grain heads pointing away from the uncut grain. With the scythe, the left hand lifts abruptly at the end of the swing, thus ending the stroke and leaving the straws together with heads pointing toward the uncut grain. Too much swing to the left, and the straws get mixed up. The scythe is lighter than the cradle, but requires more control. Grain is easier to cut than hay, since the weight of the grain heads and the height of the stem gives the plant such inertia that it has little tendency to bend rather than cut. In recognition of this greater ease of cutting, the *Larousse Agricole* for 1921 recommends adjusting the stroke and the angle of the blade to snath to regulate the depth of swath: 12–14 inches compared with 6–9 inches for hayfields and pastures.

With reaping, the support staff needs to be much larger than for haying. Thomas Jefferson hired fifty-eight people to harvest his wheat in 1795:[51]

18	cradlers
18	binders
6	gatherers
3	loaders
6	stackers
4	carters
2	cooks
57	

The fifty-eighth was the foreman or "Harvest Lord." In 1812, he recorded 15 cradlers, 15 binders, 16 gatherers (which must have included the loaders and carters), 4 stackers, and 2 cooks. The gatherers, loaders, and stackers would often be children. He describes this group as a "machine" which

> *would move in exact equilibrio, no part of the force could be lessened without retarding the whole, nor increased without waste of force. . . . This force would cut, bring in and stack 54 acres a day, and complete my harvest of 320 acres in 6 days.*

Stephens recommended the harvest be organized in units of two scythemen (without cradles), two gatherers, two binders, and one raker.[52]

So it is clear that the time spent cutting grain with the scythe is only a part of the entire harvest process. Each cut of grain must be gathered together into sheaves and bound with a simple overhand knot made of two or three of the straws. The diameter of the sheaf is limited by the length of the straw used to tie it, although old sheaf gauges used to check the sheafs size were twelve inches across. If the straws are too dry and brittle to tie, I use one or two weeds which are still green and tall from growing up at the edge of the stand of grain. Encircle the sheaf with the binder of straw or weed until the top and bottom of the binder overlap. Holding onto the very bottom of the binder, slip its top under its own base in a half hitch. Slide this half hitch along the base until the sheaf is bound tight. Bend the stiffer base of the binder back over the half hitch and tuck it under itself against the sheaf. A *woolder* for tying sheaves is a block of wood two inches square with a thirty-six inch piece of twine through a hole in the middle. One end of the twine is knotted so it is tight against the wood; the other end is wound around the sheaf and then around the twine beneath the wood block. The woolder may be reused for any binding task.[53]

Several sheaves are stacked against each other with the grain heads up to form a *stook* or *shock*. There are traditions about how many sheaves of wheat, oats, rye, and barley make a stook, but I usually lean eight or ten against each other. In the stook the grain heads will dry and harden in the wind and sun. With one large sheaf draped over the top of the stook, even the rain will be turned away, although if the weather becomes very wet, grain must be stacked and covered or brought into the barn. Thomas Jefferson had a large granary for every forty-acre field, where his grain did most of its drying. Some people like to stack their grain atop a platform that is proof against rats (off the ground on mushroom-shaped stilts), grain heads to the center, and thatch the top of the big stack or cover it with plastic. Whatever method you use, slow drying in the sheaf ripens the kernels so they will store well. Modern methods include heat-drying treatments, which often kill the germ of the grain. Taking the life away from the grain leaves no seed for next year and leaves the grain with questionable nutritional value. On the other hand, unless you have a fleet of cats, storing the sheaves loose in the barn for too long before threshing will transfer all the nutritional value to the rodent population.

Not all the grain cooperates by falling into neat bundles as it is cut, and the field may look a mess after harvest. At one time the privilege to glean a farmer's fields after harvest was much coveted because of what was left. I could go on and on picking up straws. Instead I turn the chickens or geese into the field.

Finally, the undersown clovers and grasses grow up through the stubble to give a late hay crop and a good winter cover for the grain field. I only recently discovered Thomas Tusser's sixteenth-century recommendations to do the same:

> *Corn carried, let such as be poor go and glean,*
> *and after them thy cattle, to mouth it up clean;*
> *Then spare it for rowen till Michel be past,*
> *to lengthen thy dairy, no better thou hast.*[54]

Figure 27

Tools for small grains

Sowing: *broadcast*

Sowing: *seed drill*

Threshing: *flail*

Threshing: *drum*

Figure 27

Tools for small grains (continued)

winnowing tray   riddle or sieve

Sometime in the fall or winter I take the sheaves of grain apart and separate the kernels from the straw and chaff by threshing: hitting them over the back of a chair, or banging them around inside a barrel, or jumping on them, or driving heavy-hoofed animals over them, or flailing them, or using a threshing drum. The two-piece flail, used since the fourth century, is a six-foot pole to which is hinged an eighteen-inch beater or *swingle* or *sweple* or *souple* made of hardwood or occasionally knotted rope. The swingle is brought up easily to shoulder height and brought down along its full length onto the thick pile of opened sheaves with a thump, then again, and again. The blow need not be violent nor exhausting; the grain is not beaten out, but shook out of the deep pile of straw by the blows of the flail. As with scything, there is a rhythm intrinsic to the task of threshing which makes the work easier. The leader of threshing teams coordinated the succession of blows with work songs or chants or even dances specific to threshing.

The problem of flail design is the material for the hinge. Even leather wore out quickly, so eelskin was used whenever available. Robert Johnston has proposed an excellent design to keep the swingle from getting tangled in its own hinge material (Figure 28). The heavy wire revolves freely around the notch at the end of the handle so the twine does not get too twisted.

A field before harvest may look like amber waves of grain, but threshing reveals that the kernels are but a small fraction of the weight taken from the field. The quantity of chaff and straw seem prodigious by comparison to the pile of kernels, which are one twenty-fifth of the whole by weight, and a smaller fraction of the volume.

Wheat, rye, and buckwheat thresh easily. Only the "naked," or "hull-less" varieties of barley and oats thresh easily. The hulled varieties require more work with the flail; heating and light grinding also removes the hulls.

Threshing with a flail was considered the hardest work on the farm because of the large quantities of grain that were required to feed the livestock. Mechanized threshers were greeted with a cheer. However, neither threshing nor any task is drudgery when done on a small scale; also, the work is definitely practical, and is valuable experience. Of the flail separating the seed from the hull and stalk, H.J. Massingham said, "The man who swung the flail was a person who molded the sources of life; the man who tends the machine risks himself to the inanimate."[55]

The early threshing machines to which Massingham referred were very large, noisy, and dangerous. In contrast, the foot powered threshing drum, used widely on small

Figure 28

Attachment of swingle to handle

one cubit

twine

stiff wire

handle

swingle

quantities of grains in Japan (and pictured in Fukuoka's *One Straw Revolution*), is a cylinder of wooden slats covered with wire loops rotated away from the operator by a treadle at 350–450 revolutions per minute. The straw is held in both hands over the moving wire loops, which beat the kernels out. This machine supposedly threshes six bushels an hour, with less risk of cracked kernels than in flail-threshing. However, in a series of experiments comparing threshing with the rotating drum to threshing with a flail, I found that the rotating drum performs best only when the kernels are extremely loose in the grain head. If the kernels are easily detachable and the sheaves are held carefully over the many steel fingers, the threshing drum takes the same amount of time as the flail—from a half hour to an hour per bushel—for an average of one third more kernels. Hurrying, however, decreases the yield below that of the flail. The expense of the threshing drum thus seems justified only for large quantities of grain.

Roller threshers which take in the whole sheaf between two rollers, pressing the kernels out and bruising the straw for better value as livestock feed, were once made in one- and two-person models. But no longer; the same design is used in the small and expensive gasoline-operated threshers made in this country today.

The straw remaining after threshing can be used for animal feed, for thatching of roofs, for animal bedding or garden mulch, for which it is superior to sawdust and hay, for straw rope to tie down haystacks against winter winds, or for the many beehives, baskets, and mats necessary around a farm, or for the traditional straw birthing bed. Indeed, I met several European farmers who grow grain specifically for the straw to use with their vegetables or orchards; Fukuoka's fertility program for barley and wheat depends entirely on returning the threshed straw to the fields with a little chicken manure. And straws of the threshed grain can be snipped between nodes to drink through, pleasing the children with the original organic drinking straw!

The difficulty in the past of separating the grain from the chaff, dirt, broken kernels, weed seeds, bits of stem and leaf, and other foreign matter that collect on the threshing floor or tarp is well illustrated by the many folktales where the hero or heroine is charged with the impossible task of sorting a bushel of grain in one night. Besides cleaning by hand, the traditional method of separating grain from waste was to wait for a favorable breeze, then throw the mass of grain and other matter up into the air; the chaff and dirt were blown to the side while the grains dropped back into the winnowing tray. The chaff was saved to fill mattresses, store sausages, etc. To separate the grain from the bits of stem, a series of *riddles* or sieves was used through which the grain kernels dropped. Handpicking of remaining impurities followed, making grain cleaning the most time-consuming task of harvest. Washing the grain—rinsing then drying on trays—floats away the remaining chaff and cleans off any dirt. Even so, bread must have concealed other ingredients besides grain meal.

An early improvement was the winnowing fan, used to create a draft on a windless day. Odysseus's last task was to travel inland with an oar over his shoulder until he should meet someone who mistook it for a winnowing fan. Some small-scale farmers today use an electric fan with the same effect. The winnowing sheet was flapped for a breeze a great deal in the United States prior to the nineteenth century; three-quarters of an hour were required for every bushel of grain. Studies in 1898 concluded that over twenty-six hours were required to thresh twenty bushels with a flail, winnow with a sheet, sack the grain, and stack the straw,[56] much more time than modern farmers feel they have for growing their own grains.

The rotary winnowing machine with crank handle, in use in China by 40 BC, was brought to Holland, and then to Scotland in the late eighteenth century, a lag of fourteen hundred years. By 1794, it was used by every husbandman and made by every country carpenter in Britain.[57] Recent improvements have been in the types of screens used. In the hand-cranked version pictured in Figure 27, threshed grain is dumped in at the top, and a hand-crank turns a fan as well as jiggling three screens. The screens are selected from over a hundred for the specific sorting task, experimentation which I find very enjoyable. Particles that are too large are directed out one spout, too small out another, chaff is blown out a third, and the grains are separated into two sizes. The output can be twenty to thirty bushels an hour, less if there is a great deal of straw with the grain.

Storing in the grain bin or granary requires dryness. The bin should be elevated off the floor so air can circulate beneath it. Several two-by-fours can be set vertically in different places in the bin. When the bin is full, the wood is removed to provide passages for air movement. Protection from rats is obtained with heavy screen or solidly constructed wooden bins. The possibility of insect infestation can be eliminated with small quantities of grain by setting portions in a home freezer for a day, or, with larger quantities, by storing outside in the winter cold. Dry ice (carbon dioxide) is feasible for killing insects in the grain with large or small quantities. These treatments do not kill the germ in the kernel as do high heat and strong chemicals.

As for grinding and cooking grains, there are many books on the subject. It should be said that the whole grain is a natural storage container in which the nutritional qualities and seed viability are kept intact. Seeds unearthed from the tombs of the Egyptian pharaohs have sprouted when moistened, and wheat has been found to have a germination rate of eighty-five percent after thirty-two years in storage.[58] This protection is destroyed when the kernel is cut in any way; then the contents begin to lose their potency and the oils turn rancid. It is best to keep the grain whole until shortly before it is used. Then the full life-giving power of the staff of life can be utilized.

I wanted to know just how my homegrown wheat measured up to other available sources so I designed a simple experiment to test it. Into each of two bags I sifted six cups of a commercial brand of "whole wheat flour" (though it is called whole wheat, it does not have the full germ since this would shorten its supermarket shelf life). Into two more bags I sifted six cups of hard spring wheat from North Dakota which I had ground into flour that morning. Into the last two bags I sifted six cups of wheat I had grown and harvested myself, and ground into flour that morning. I randomly assigned the letters A to F to the six bags, then delivered them to a professional whole wheat baker with instructions to treat all six bags in the same manner. Using a simple recipe for whole wheat bread by the sponge method, he baked six loaves that day. I arranged taste tests that evening and the next two days for several individuals. The judges unanimously awarded last place to the commercial flour. Distinctions between the other four loaves were more subtle, most judges having good things to say about one individual loaf or another. The baker's favorite was a homegrown loaf: it rose most nicely, sliced the thinnest, and had the best flavor. The majority of opinion was for the homegrown bread. When the identities of the flours were revealed, most judges groaned, saying, "I guess I'll have to grow some wheat next year." I must confess I was surprised: "just as good" would have been adequate, and "superior" was an unexpected confirmation of my labor.

BRINGING THE HARVEST HOME

Your hay it is mow'd and your corn is reap'd,
Your barns will be full and your hovels heap'd.
    Come, boys, come,
    Come, boys, come,
And merrily roar out our harvest home.
       —JOHN DRYDEN AND HENRY PURCELL, *KING ARTHUR*, 1691

When grains were harvested by hand, the fields could be cared for in smaller parcels, and with more careful scrutiny than is usually possible today. Men, women, and children who worked together around the center of a farm, community, or neighborhood—home in its broadest sense—would come together in work at harvest time. The participants might act in different occupations for the rest of the year; but these could be put off. The harvest could not.

The link to the land and to the harvest task was also shown in the form of payment. The part-time laborers could take home a portion of the crop, or be paid in other forms of the farmer's, and therefore the community's, wealth. For example, Thomas Jefferson set the "proper allowance" for six long days of work from each of his fifty-eight wheat harvesters at "4 gallons whisky, 2 quarts molasses, 1 midling (i.e., side of bacon) besides fresh meat per day, with peas." This was in addition to a portion of the crop, also given in payment. In thirteenth century England, it was customary for the lord to grant his mowers as much hay as each could lift on his scythe at the end of the day; at the end of the season, they were given a sheep to catch and share.

Shared meals deepened the relationship between the harvesters. H.J. Massingham recorded eight meal breaks in the harvest day:

> Forebit and breakfast,
> Rearbit and dinner,
> Nunchins and crunchins
> And a bit after supper.

The confirmation of the community family through work and the close bond to the earth through agriculture have been felt since the beginning, and it is not merely nostalgic to bemoan the loss of the relationships in industrial societies. The loss of the traditional forms of cooperative labor means it is more difficult to find the opportunities for companionship in work. Approaching hay, weeds, or small grains with a scythe is one such opportunity. To the hard-headed realists which we have all become, a scythe or other tool must first make sense economically. Beyond that, we are free to discover the other benefits—physical, emotional, intellectual, and spiritual—which the use of the scythe permits. Exploring old customs can help in the articulation of these benefits.

One old custom has been the portrayal of Death and Time wielding a scythe. (Figure 29):

> See how Death preys on humane Race;
> Out with his Scythe the Tyrant goes,
> Great Multitudes at once he mows.[59]

If the scythe is the Grim Reaper's favorite tool, was it considered to be a terribly dangerous weapon? No. Sharp as they are, scythes could never be used in combat unless the blades were reattached to extend straight out from the snath. Such a connection is tenuous, and did not seem to work well. Stories that the Assyrians attached scythes to their

Figure 29

Death reaps with the scythe

chariot wheels to mow the enemy down are a mistranslation: knives and swords were used for this purpose, while scythes could not be. In fact, the scythe is the very symbol of peaceful, settled farming of the land. In *The Book of Isaiah* (2:6), the nations that have given up war "shall beat their swords into ploughshares and their spears into scythes." For John Milton also, the scythe represents the peaceful pastoral life:

> While the ploughman near at hand
> Whistles o'er the furrowed land,
> And the milkmaid singeth blithe,
> And the mower whets his scythe.

Every author cited in the section on technique has referred to this sweet ring of the blade when sharpened with a stone.

Then why, if the scythe's other connotations are so benign, do Time and Death wield a scythe? Because mowing has a rhythm so strong and compelling that it feels, like Time, unstoppable. There is a correspondence between the inexorable laying down of the grass and the inexorable laying down of our lives. "For they shall be cut down like the grass" (*Psalms*, 37:2). This compelling rhythm is stronger than with any other tool I have experienced. Time could have had a hammer and knocked heads about, or a saw, or . . . All these other tools conjure up more grisly images for Death than does mowing and reaping, acts linked to the earth's seasons in parallel to the thriving and passing of a person's life. And in the rhythm of creation, Death's reaping is also linked to the hope of spring's recreation of life.

> And nothing 'gainst Time's scythe can make defense
> Save breed, to brave him when he takes thee hence.[60]

So grain, and the tool to harvest it, have been linked to death and burial, conception and birth.

Customs that deal specifically with the end of harvest, with bringing the fruits of the earth home, have varied from region to region. A common experience is the feeling of being finished, released from toil, when the last load of grain or hay is brought back to the barn, amidst shouts and songs:

Harvest Home! Harvest Home!
We've ploughed, we've sown,
We've reaped, we've mown,
We've brought all the harvest home.
Harvest Home! Harvest Home!
We want water and can't get none!
[whereupon the singers were drenched with a bucket of water]

Finished, but not released altogether. Since ancient times a common thread in the customs of farming peoples everywhere has been the belief that the last sheaf embodied the living Corn Spirit ("corn" as the staple grain of a particular area). All the harvesters cut the last sheaf of corn together—the single remaining stand of vertical grain in a field now flattened by the harvest. This moment was extremely potent; a stranger passing by the field at this time was scowled at, while the mowers loudly whetted their scythes, or he was approached threateningly with scythes, or he was tied up in a sheaf and motions were made to cut off his head. Thus the stranger was treated as a disturbance of an important ritual, and also as an embodiment of the spirit of grain. Often the very last ears of grain were plaited together, and occasionally dressed as a Corn Dolly to preside over the farm table at the Harvest Supper and to be kept until the next harvest.[61]

I, too, have felt the crackling concentration of energy in the last upright bunch of grain in a field laid flat by the scythe and dotted with the piles of tied sheaves. Where does that energy go when the last bit is cut? I do not know, but I understand the impulse to treat the last sheaf a special way as if the special energy stayed with it, and I do treat the last sheaf of my harvest in a special way.

The welfare of the crops was to the mind of the ancients assured by the Goddess of Agriculture—Demeter to the Greeks, Ceres to the Romans. In Figure 30 is pictured an eighteen-inch high votive mask of Demeter used in ancient Greece with the prayer: "May it be mine, beside Demeter's altar to dig the great winnowing fan through her heaps of corn, while she stands smiling by with sheaves and poppies in her hand." In more recent times, the Corn Dolly has held the Grain Mother's assurance through the winter of spring and new crops—the Dolly *is* Demeter until the Goddess returns to the meadows in the spring. The Dolly is therefore given due respect at the autumn celebrations of plenty. "That great and marvelous mystery of perfect revelation, a cut stalk of

Figure 30

Demeter

Figure 31

The harvest

Figure 32

Cradling wheat

grain,"[62] was the center of agricultural ceremonies for thousands of years before Christ. And, as bread, grain has continued its mysterious, religious association with the body of Christ ever since.

Our celebration of Thanksgiving late in November has its roots in this same tradition, though it has lost its attachment to the season's rhythms, to the feeling of being done with work in the fields and gardens, to the feeling of barns and homes bounteous with the fruits of the earth, providing nurturant protection against winter. Now the dimly remembered strings of the harvest melody are plucked by the sale in gift shops of plaited Corn Dollies, mass-produced by an expert hand. But the year-round availability of harvest trinkets is not the point living on the land. The embodiment of the Corn Spirit need not be a handsome, saleable item. Its purpose is to intensify in a symbol the communion of the farmer/gardener, through agriculture, with a little piece of the earth. For me, the painting of the harvest by Emile Bernard in Figure 31 evokes most powerfully the emotions of mingling with the Corn Spirit, and drawing up the energy for our sustenance out of the earth.

The attachment to earthly rhythms may be lost in the modern industrial age, but is hardly unrecoverable. While we no longer know a dozen traditional harvest songs to sing around an overflowing Harvest Supper, the experience of relating to the earth in a profound way is always available. It is just outside the door. The Findhorn Community in Scotland has even brought the pursuit of greater awareness to more articulate levels in work with plants, and with the seldom-seen spiritual beings that surround them.[63] I do not believe this just because they have written a book about it. I am practical. When I talk to the grasses, weeds, and grains, and spend time listening in return, does it change the nature of the crop? Does it change my experience in sowing, reaping, and eating? I will not argue whether this is fanciful superstition, or a dutiful and necessary recognition of the elemental kingdom. At the very least, I know, the old adage is true: "Farmer's footsteps are the best manure." It is my desire on this land, in the context of this small farm, to know the best use of my body, and to move with the scythe as an extension of my body—in relationship with the grass, and in a team of mowers—in the process developing my own communion with the divinity that creates the task and all its parts.

THE SCYTHE MUST DANCE

THE PRACTICAL USE
OF THE SCYTHE

Peter Vido

DURING ONE OF MY WINTER JOURNEYS TO AUSTRIA, LAMBERT SCHOISWOHL, A retired scythesmith, told me "Die Sense muss tanzen" (The scythe must dance). In the summer of 2000, Manfred Atzmüller, the mayor of Rossleithen and a mower himself, organized a brigade to trim some of the roadsides in the village. And so I had a chance to mow along the side of Mr. Schoiswohl. At 72, the extremely steep slopes did not hinder him.

On December 24, the most celebrated day in Europe, he finished decorating the Christmas tree, stood back to revel in his last task—and was a mower no more . . .

The following poem is dedicated to the memory of Lambert Schoiswohl and other "old" men with gifts in their hands.

—Peter Vido

The Old Man's Gift

"The scythe must dance," the old man said
"Yes, I can imagine it in my head;
But is this step first or that instead?
There's not a guide which I have read.
Will you teach me all before you're dead?"

"See what I do—watch closely now!
The dance you'll learn—I'll show you how."
The young girl smiled and gave a bow
As she thought of a fragrant hay-filled mow
With leafy feed for her only cow.

Together they met in the meadow each morn
To dance before the day was born.
One taught the steps, one learned the form,
Stone often slipped from water-filled horn,
Blade slicing smoothly grass and corn.

Gazing upon neat swaths with pride
And gratitude she did not hide,
thanked the teacher at her side
For a skill which makes the heart so blithe—
The old man's gift before he died.

FAYE ABEND VIDO
JANUARY 2001

Introduction

EVERY VERSION OF WHAT WE MAY, AT A CERTAIN POINT IN LEARNING, REFER TO AS "complete" is, of course, only relative; the path to knowledge is infinite. David Tresemer took the scythe related research, in his own words ". . . as far as it could go." At the time I crossed paths with *The Scythe Book*, it was exciting to learn that someone *had* indeed written so much about the tool with which I was just then falling in love. Today I am dismayed at how *all* written material I have managed to find thus far on how to use the scythe is sorely lacking in depth.

Nevertheless, the poetry and old accounts of mowing which Tresemer included in his book, along with the metaphysical undertones of certain passages, eventually played a role, perhaps a significant one, in inspiring me to continue the search.

The tool itself I could readily appreciate on its own merit. Growing up in Slovakia, albeit as a town rather than country boy, I saw the scythe being used very frequently. It must be an energy-efficient method of harvesting substantial areas of forage, I reasoned, or the socialist system, all its various flaws notwithstanding, would not so depend on it. Until the era of the Communist regime ended and brought along certain kinds of "freedom," practically all roadside ditches as well as grassed areas within town limits were trimmed by elderly men with the scythe. [See note 1.]

Coincidently, that very same summer when one of our farm apprentices who came to learn how to work with horses, brought along his copy of *The Scythe Book* (which I hitherto had not seen), my parents from Slovakia were also here on an extended visit. I was teaching the apprentice how to use, among other implements, the sicklebar team mower, while my father, with an old scythe I once bought at an auction sale for a few dollars, went about trimming the areas the horse-drawn machine could not reach. By a stroke of luck the scythe consisted of a European pattern Austrian-made blade with an elevated tang mounted on a better-than-average light aluminum snath. My father was not a man of many words, and so *The Scythe Book*'s philosophy nicely complemented the events taking place. I tried out the scythe and liked it instantly, though did not begin to use it seriously until the next season. A turbulent storm had made such a mess of a couple of acres of buckwheat that any mechanical cutter would have proven useless in harvesting it. With the scythe I could follow the trail of the wind, snaking through the field in whichever direction the stems required. It was empowering, and I cut every bit of it plus several more acres of well-standing grain that fall.

Next summer, while still cutting 50 to 60 acres of fields with the horses, I followed through those little places in my father's footsteps . . .

By then, however, the blade needed very frequent honing. Having lived as a farmer/homesteader for 15 years, the subject of sharpening in general was not new. I was learning to use axes for tasks other than just splitting firewood and also knew that it made a difference, even to a team of large drafters in good condition, if I pulled the knife out of the sicklebar mower each time I returned to the barn, whether I had mowed an acre or half a dozen that day, fixed all minor damages, and sharpened it again.

The cold shaping of an edge, however, was different. Back in 1977 my old farrier instructor's first lesson was: There are only two transgressions for which a blacksmith can go to Hell. The first is hammering cold steel; the second, not charging enough. Money meant little to my father. He consequently never instilled in me that it was an honoured means of exchange for the expending of one's energy, so I was already on shaky ground . . . Going to Hell was not yet on my agenda; I was having too much fun mowing by hand, among other things.

Yet an old Slovak axiom states: "Peening is to the scythe blade what bread is to the mower." David Tresemer endorsed it. So risking that gate into Hell, I gave peening a try. This is when I learned that I was of less than "ordinary intelligence" [see note 2] and naturally slow with my hands. I could not do it in five minutes (I still can't), never mind doing it well. That first blade took some abuse.

There must be more to learn, I reckoned, and hopefully, somewhere, more detailed instructions than *The Scythe Book* contained. Endless hours of letter-writing to Europe bore but little fruit. My search eventually took me, several times, to Austria, the cradle of scythesmithing, with side trips to surrounding countries. It helped my learning immensely, though that long sought, detailed "scythe user's manual" is still but a dream.

Perhaps it lies, dust-covered and forgotten, in some computer-less library—"The Zen of Scything"—written by a real master of this ancient tool. Until I discover it, or a true authority is found to take on the task of writing such a book, I humbly submit to serving as a poor substitute and hereby offer you a brief summary of what I have learned thus far.

What David Tresemer did for all of us was that he fanned the embers of a dying fire. For this I am grateful and herewith add more fuel to the flickering flame. In doing so I hope to increase the probability that this most amazing and multi-faceted of the tools I have been privileged to use will gain broader appreciation. Not as one of necessity in the hands of toiling peasants as it was in the past, but rather as a tool for the visionaries of a better tomorrow.

I

Clarification of Terms

SOMEWHAT RELUCTANTLY, AND IN AS FEW INSTANCES AS POSSIBLE, I CONTINUE referring to the subject of this book as the "Austrian style scythe." It is convenient to do so, because David Tresemer—or someone before him—established the term in North America and many readers at least partially know what is meant.

However, it is a misnomer, which does not acknowledge the fact that this type of blade has also long been made in other European countries. In the respective languages the tool is merely referred to as a *scythe* (and this is my preferred term as well). Sometimes, if the country of origin is known and it is desirable to mention it, the blade is given the prefix of "Austrian," "Russian," "German," etc. to emphasize its reputed quality relative to another.

Apart from Austrian-made blades, others of the same "style" have been imported into North America for well over a century. Most were models not used in Austria at all, even if made there, and some were forged in Germany, France, Sweden, the former Yugoslavia, or perhaps elsewhere.

It would therefore be more accurate to call all of these collectively "the blade style of Continental Europe" as distinct from those made and used in the British Isles and North America. What differentiates them more significantly than their physical shape is the property of *tension* incorporated into the steel during the process of manufacture, which allows the former to be lighter in weight, yet still retain its shape in use. The British and American-made blades were not tensioned and therefore, *relative to the task* for which each respective model was made, had to be heavier.

Yet I do not refer to the latter as "hard stamped"; in fact I do not use that term at all. Having held in my hands many samples of the "American Pattern," I detected *only one*, which was distinctly stamped. It was a brand new blade of old inventory I found in an Amish-owned hardware store in Indiana with a label "True Temper," Made in USA. How many European scythemakers ever stamped blades I do not yet know, though I think that it would have been very few. The best of them would not have dared to mar their reputation in this way. The chief technician in the Austrian factory which presently supplies most of the stock sold in America personally assured me that they had never stamped a blade of any kind. (Hay and straw knives, swords and, of course, sickles, have also long been forged there.)

Using "stamped" as a blanket term for all non-"European style" blades discredits the skill of all those British and American blacksmiths who did hand forge their creations.

The concepts of hardness and "peenability" of these flat blades I discuss elsewhere in this text.

To a born European, for reasons irrelevant here, "America," as a reference to that big place over the ocean, usually includes the USA and Canada. After living on this continent as a Canadian, for 33 years now, the associations remain to me, for the most part, unchanged. Thus, instead of specifying USA or Canada, I use "America" as a term inclusive of both.

Sharpening, in technical jargon, is a two-step process consisting of i) beveling and ii) honing.

My attempt here is, therefore, to avoid the use of the verb "to sharpen" (the blade) whenever I am talking specifically about either "beveling" or "honing."

Thus, for example, I will not write: "In the field, during mowing, the blade must be *sharpened* every five minutes." Instead it would read: ". . . honed every five minutes." *Whetted* would also be correct. Because honing, at least in my mind, is more indicative of the use of finer abrasives, which I advocate, I prefer this term and use it more frequently.

It may be pointed out that "honing" also bevels the edge, albeit to a lesser degree. This is true, and I will endeavor not only to explain that it is so, but also to emphasize that the creating of the often-unnoticed little bevels (more properly called "secondary bevels") during honing is, for the most part, unintentional and undesirable, though in freehand practice unavoidable.

Another term for beveling is *shaping* (the bevel of) the edge. The blades discussed here are best *shaped* (i.e. beveled) by the process referred to as "peening."

Alternative terms by which most scythe users might describe this activity are *hammering* or *thinning*. Both are correct in that a hammer is used and the edge will inevitably become thinner. However, I prefer *shaping* because it alludes to the importance of a more exacting geometry. This will be discussed more thoroughly later in the text.

Below are groupings of terms with essentially the same meaning, which I use alternately because in some instances the slight variations portray better what I am trying to say. In any one group, the term(s) in italics is my favored one.

Sharp, razor-sharp, well-honed, *keen, cuts with ease.*

Not sharp, blunt, dull, relatively useless, edge reflects light, *lacks keenness.*

Peening, hammering, thinning, *beveling, shaping.*

Finishing, whetting, *honing;*

Whetstone, finishing stone, *honing stone* or simply *stone.*

(Hammer) blows, strikes, *taps* (because it indicates more gentle and careful use of the hammer).

"Serrations or sawteeth;" feather edge, wire edge, *burr* (all refer to a condition when minute particles of steel protrude over the edge before they are removed with very fine abrasives or break off in use).

Also, I often substitute "scything" with *hand mowing*, or just *mowing*. Please do not ask me to explain why. This book is about *scythes*. Whenever I indicate mowing with the tractor, horses, or by other means where a machine is used, I will make that clear.

In some instances, one may need to pay attention to the difference between "v" and "w" in the words moving and mowing (though mostly the distinction will be irrelevant). When I move, as in ". . . across the slope, in the direction of the wind . . . ," I am moving while mowing. In these pages I seldom move without the scythe, or at least its blade, close to my hands . . . So, whenever I write *this tool*, or just *tool*, unless specified or used as a reference to other tools in general, I mean the scythe.

II

Fitting the Snath
to the Mower

THERE ARE TWO ADDITIONAL TERMS I HAVE "MADE UP" FOR THE PURPOSE OF referring more accurately to what I view as two versions of this tool. The terms are: *field mowing scythe* and *trimming scythe*.

Related to these, of course, are field mowing snath and trimming snath, respectively, used in field mowing and trimming. (By *field* I mean a larger area, usually more open and without obstacles; I may likewise describe it as a hayfield, meadow or a lawn).

It is a somewhat arbitrary distinction, because during the mowing of a large field or lawn-like area there may be occasional need for a short trimming stroke, as when going around a boulder, large anthill or a lawn ornament. Conversely, while working in an orchard where the trees are closely spaced and for which the *trimming scythe* is generally better suited, one could, in cutting perhaps half of its total square footage, use a longer blade and/or snath (i.e. the *field scythe*) with advantage.

Even though each tool can partially replace the other, I am convinced that the virtues of the two distinct versions would be readily appreciated by anyone with a little experience and diverse mowing tasks.

While I can, for instance, accomplish a lot of trimming using a field snath, in doing so I would be continuously aware that it is more cumbersome, less accurate, and causes my arms to tire sooner.

Figure 1

[A), B), and C) for my size, 5' 10" with bare feet]

(A) *Trimming Model-Body: Alder; Grips: Beech; Weight: 2 lbs 4 oz.*

(B) *Field Model-Body: Aspen; Grips: Apple; Weight: 2 lbs 10 oz.*

(C) *Compromise Model-Body: Ash; Grips: Spruce; Weight: 2 lbs 14 oz.*

A. Trimming Model
Body: Alder
Grips: Beech
Weight: 2lb. 4oz.
20" 36" 54"

B. Field Model
Body: Aspen
Grips: Apple
Weight: 2lb. 10oz.
22" 38" 58"

C. Compromise Model
Body: Ash
Grips: Spruce
Weight: 2lb. 14oz.
21" 37" 56"

Figure 1

[Continued D) Fairlight's, E) Ashley's]

(D) Children's Model (for 10-12 year olds [4'8"-4'10" tall] using blade less than 14 oz.)-Body: Alder; Grips: Ash; Weight:] lb 6 oz.

(E) Children's Model (for 6-9 year olds [4'2"-4'6" tall])- Body: Alder; Grips: Spruce; Weight: 1 lb 8 oz.

D. Children 10-12 yrs.
   4'8"-4'10" tall

Body: Alder
Grips: Ash
Weight: 1lb. 6 oz.

18"   31"   47"

E. Children 6-9 yrs.
   4'2"-4'6" tall

Body: Alder
Grips: Spruce
Weight: 1lb. 8oz.

17"   28"   44"

The following is the result of my making and using dozens of snaths of various shapes and grip positions. Even though the experimenting continues, I have concluded that the measurements suggested here work best for me and that two inches one way or the other may affect the experience significantly.

Figure 6 on Page 12 of *The Scythe Book* can be used as a reference to the procedure of fitting, though I would like to be more specific. In my experience of fitting snaths to people, nearly all, when asked "Where is your hip joint?," indicate their waist. I want to emphasize that the spot we are looking for is approximately 5 to 6 inches below the waistline. A tailor wanting one's hip measurement would lay a tape exactly over this area. Anatomically, the hip joint's bony protrusion is referred to as the greater trocanter.

I suggest that the measuring be done in the footwear similar in heel height to what the mower wears while at work.

A few snath samples (including specific measurements) are depicted in Figure 1.

The lower grip of the trimming scythe (the snath with the blade mounted) held along the side of the body (as pictured in Figure 6, Page 12 of *The Scythe Book*) should reach to the above discussed point on the hip, *never below it*. The distance between the grips should be one's cubit (as defined on Page 12 of *The Scythe Book*); *no less*, but also not much more. Figure 1 is an example of such a fit for my proportions. I will put up with two to three inches more than this only if I am using a snath the top end of which drops down as in a) or c) in Figure 1. On a *straight model of snath* this extra length would require that I hold my left arm higher than I like to, because it predisposes the building up of tension under the left shoulder blade.

For a field snath I like the lower grip two inches higher, as in sample b) of Figure 1. The distance between grips should be one's cubit plus two to three inches. In field mowing the blade is moving further away from my feet, with the snath on a lower angle to the ground. This allows my left arm to be also lowered and thus stay in a more comfortable position. The extra inches, therefore, provide better leverage without paying the price of the left shoulder's discomfort. For the wide field movement, especially in heavy grass, one's cubit distance between grips is not enough, unless the mower has very strong arms.

Should you choose to permanently use only one snath for both types of work, a "compromise model" may do.

It needs to be pointed out that many people are not built proportionately the same. Some of us have extra long legs (thus the hip joint is further from the ground) or extra long arms (so the cubit is longer). A custom-fitted snath for someone 5'10" tall (even if he mows barefoot) may not be of the same dimensions as the samples a), b), and c) in Figure 1.

Children need, or seem to do better with, snaths slightly longer in proportion to their bodies than the specifications for adults as discussed above. Sample d) is one of Fairlight's *field* models. She is eleven years old and 4'8" tall, i.e. 14" shorter than myself; note the difference between her snath measurements and mine as in sample b). At the same time, children's flexibility affords them more leeway, and sample e) was one of the snaths she could also use this past season, especially for trimming, albeit with a *steeper tanged blade.* (This topic will be discussed later.)

Her sister, Ashley (4'3", age 8) can both trim and field mow, again using different blades, with sample e). But because she has a naturally wider reaching style of movement and likes to mow at least as wide a pass as her sister, the most recent snath I made for her is almost the dimensions of sample d). Note that e), despite being shorter, is slightly heavier. The extra weight is above the lower grip to help *balance out the blade*, which for children is even more important than for adults.

III

How Tough Is a Scythe Blade?

I AM NOT PARTICULARLY GENTLE WITH ALL OUR BLADES AND CONTINUE TO BE awed by how much they can take. They are confronted with rocks frequently. Usually I survey the damage to the rock as well as the edge. Often enough it happens that a piece of the rock is missing and nothing from the blade. At times I have to look carefully against the edge for the light reflection to even know which portion contacted the rock. I have run into wire fences and steel ground rods with these blades and would indeed dare to mow a junkyard clean, weaving in between all the car parts strewn about

While at the Common Ground Fair in Unity, Maine, I was discussing the "toughness" aspect of scythe blades with Marc Harmon, a man who had spent most of a day around our demonstration area asking "above average" questions. We happened to be squatting by the back of our '76 Chevy LUV and I had in my hands Fairlight's 24" Turkish pattern blade. To emphasize that it is stronger than it looks, I held it by the tang and drove the point several times against the truck's steel bumper. Amazed, Marc inspected first the blade, which bore no marks of the combat, then the bumper which showed definite indentations.

Right away he wanted to buy that very blade, though I assured him that there was nothing special about it, and every other one of these Austrian blades is made to the same specifications. He still insisted, so I let him have it.

Thinking it was rather funny, I related the story to Scott Howell of Blue Hill, Maine, who came by soon after, and with another used blade, that happened to lay there on the table (a 22" Greek pattern), showed him what I had done. Then I walked away to talk with someone else. On returning I learned that Scott, likewise, convinced my son to sell him that little blade, which Fairlight and Ashley had used, among others, that summer. I found this even funnier. You see, Scott is built like a blacksmith of pioneer days when all the horses, willing or not, had to be shod . . .

When he came to our workshop two years ago he brought his own scythe, the battle wounds of which were a testimony to his strength. Subsequently I made him a snath of ironwood (hop hornbeam). There are blades more suited for extra hard-working people. They would not be one of the above-mentioned models, which are very narrow and among the lightest in weight still produced in Austria today. Neither of them have "stone-points." They even look too delicate for "real work," I have been told.

This is an illusion. Those two blades and others just like them have cut a lot of tall, dense grass here, as well as many mighty looking burdocks, two-year old green poplar trees, clumps of mature goldenrod, raspberry canes and so on. . . .

They were made by the Styria scythe factory in Rossleithen, Austria; whence also come most blades on the American market today, including the "flat ones."

Certain written sources erroneously refer to the "Austrian style" blades as being made of "mild carbon steel" or "soft steel so they can be hammered." I reckon there is little that is mild about a blade which was forged out of steel with 0.8 carbon content and then tempered to the hardness approximately equal to that of a good quality ax—unless one's definition of "mild" comes from some, to me, unknown classification tables.

Styria's "Austrian," "American" and all other patterns are made of the same steel, shaped under the hammer and tempered very close to the same hardness, which today is 47 to 48 Rc. (Rockwell scale).

I hope that my driving the points of scythe blades into truck bumpers will not encourage others to do the same. Let me add here that I have never bent or broken one and that, a few years ago, I would not have dared to do this.

IV

Blade Selection

SCYTHE BLADES ARE STILL MANUFACTURED FROM 12 TO 50 INCHES IN LENGTH, IN many distinct patterns, and still more models of each of these. The lengths from 24 to 32 inches are probably used most frequently worldwide. Yet to decide which is the *most suitable* single pattern, model and length for the average American user of today is, I feel, an impossible task. The issue should, however, be given more consideration so the decision on the level of compromise can be made somewhat knowingly and *collectively* by dealers as well as the customers.

The shortest and the longest blades represent the extremes from maneuverability to the potentially largest area of grass that can be cut with one movement. To limit the customer's options to but one blade is unfair. Whatever way it is justified, it will not best serve most mowers under all conditions. Further, it propagates ignorance in that most customers are not told, and so are unaware that options do exist. Conversely, for any mower who is offered several blade lengths to be content with only one depreciates the scythe's potential as well as his own. It is poor economy from the standpoint of time even if one's pleasure in using a well-fitted and suitable tool is not taken into account.

In spite of my barefoot mowing, as I sit here searching for an analogy, *shoes* of all things come to mind. We all *could* get by with one pair of them. In comfortable hiking boots we could go to church or a dance year round, climb snowy mountains, stroll tropical beaches or jog the city pavements. Why we don't, each individual no doubt has personal reasons for. Yet, from the perspective of one spoiled by having so many different scythe blades, I would find living with one pair of shoes easier, or more acceptable, than with but one blade.

Please realize, therefore, that in my effort to help you decide which is the one "ideal" blade, any way I write it down, I could argue with myself.

The lengths available at this time in America regarding the purposes as discussed in Chapter 2 might be divided as follows:

20, 22, 24 inches — trimming
26 and 28 inches — compromise
30 and 32 inches — field

This is not to say that one could not cut a 10-acre field with a 20" blade or trim a yard with a 30" one.

A novice in a confined area will find a short blade much easier to maneuver and so be more accurate in getting close to tree trunks and fences etc. It will also be less over-whelming to hone and peen.

The long blade's ability to cut off more at a stroke is encouraging and increases one's sense that it is worthwhile to spend the time in a field or large lawn with a simple hand tool. The virtue of longer blades is in allowing a deeper *forward bite* into the grass, *regardless of the swath's width.*

Those who consider cutting several acres of hayfields would find a true field-mowing blade notably advantageous.

Having two lengths of blades at my disposal, instead of only one, I would find far more than twice as suitable. In space-constricted areas within a garden, yard, orchard or around obstacles, I make better time and do a neater job using a 22" blade than a longer one. It is difficult to explain on paper when, under the above conditions, I would choose a 24" to do the job a bit faster or a 20" for more accuracy and tidiness. I may at times take even a 26" and decide to trade off more maneuverability for the time element. The choices are not hard to make by a quick visual assessment of the site once you gain some experience.

A 22" may be my favorite of the "short lengths," and highly recommended to those who would later consider an addition, but like to "start small." (Use it also on the trim-ming-size snath!)

A good length to complement this one would be 28" for the cutting of more open areas. Those who, like me, consider the scythe a serious grass-cutting tool may add a 30" model instead. Nice two-blade combinations are:

22" and 28"
24" and 30"
20" and 26" — for children or "small area specializers"
28" and 32" to 36" — for relatively large areas with a smooth surface.

If I had to part with all but two blades, I would keep a 22" and 34". The "general purpose compromisers" may be satisfied with a 24" or 26" if they do predominantly trimming work or a 28" if most of their grass surface is relatively open.

BUSH BLADES

David Tresemer's reference to "bush or brush" blades is limited to one sentence. In a way I feel like leaving the issue at that because the bona-fide bush blades and the technique required in their use are not well suited to the kind of "T'ai-chi-like" mowing which I wish to help the readers experience. They were developed for the strong mowers of old who had a lot of work for such a tool. However, if you feel the need to acquire one, it does not need to be long; it is used with a much shorter movement, often directed at a single stem or a handful of them. A length of 20" is plenty and 18" or even 16" will suffice. In most cases the weight difference will be appreciated.

There have also long existed two types of blades that span the difference in purpose between sturdy bush blades and the light grass blades. These are the so-called "Forest Culture" and "Ditch" blades. I find much more frequent use for them than their heavier cousins. Moreover, there is little I encounter on this farm that a grass model with an appropriately shaped edge will not take care of, if used with good judgment.

V

Sharpening Principles

EVEN A SUPERB QUALITY BLADE AND ONE PARTICULARLY WELL SUITED TO THE task will prove of little use if its edge is dull. The importance of the mower's sharpening skill can hardly be overemphasized. In order to avoid being caught by the relativity of the term "sharpness," we can simply say that a sharp blade is *one which cuts with ease*.

At a certain early stage, believing I already knew what "sharp" meant, I dared to share my "knowledge" with some of my scythe-using friends. But perhaps the following season I came to grasp other related details and consequently doubled my average hourly accomplishment, or cut each acre with twice the ease. This phenomenon is by no means limited to scythe blades and I think that a seasoned butcher, old-time barber or woodworker would agree. Many years after you already regard yourself as an experienced mower, you will still be learning to sharpen—and so will I. Consider, therefore, the following only as an attempt of one apprentice to share with another.

The cliche "razor-sharp" is often thrown around without the understanding that, with reference to many blades other than razors, the analogy is somewhat meaningless. That sharpening is a process of thinning metal edges is more or less understood by anyone using cutting tools. The casual user, however, gives the *specific shaping of bevels* too little consideration. As a result many tools like knives, axes and yes, scythes with relatively sharp or even "razor-sharp" edges do not perform well. To express this alternately, it can be stated that a thin edge does not always mean "easy cutting" and thick (relative to the other) is not necessarily dull.

Most of the present day woodworkers who depend on hand tools understand this, as did, for instance, the best of the lumberjacks. Some were renowned for their skill in *shaping the bit*. Consequently, their axes took the largest chips out of tree trunks, yet had no tendency to "bind." Many of these men could not explain it well in theory, nor even knew the specific degree of angles to which they ground and honed their tools. What they did know was how to make their axes *cut with ease*. (Dudley Cook, by the way, explains the pertinent principles in *The Ax Book*, the only comprehensive text on the subject of axmanship I have found to date.) [See note 3.]

Because we cannot view a scythe blade's profile (unless a section of the blade is cut off), let me use axes as an example to illustrate the concept of shape. The geometry of their fore sections and edges is large enough to be looked at easily.

It is not difficult to put a hair-shaving edge on almost any ax, or even a splitting maul, with only hand tools and a few minutes of time. [See note 4.]

A splitting ax with such an edge will be useless as a tree-felling tool. To sever wood fibers, in addition to the keenness, it would need a more penetrating profile, such as is afforded by the innate shape of a felling ax. Likewise, a scythe blade that may shave hairs is not necessarily *also* one that, in some mowing conditions, will cut with ease, not for more than a few strokes in any case. On the other hand, it could be less keen than "razor-sharp," but *suitably shaped*, and it will perform well. The same can be said of the felling ax. However, "suitably shaped" and "thin" are not necessarily synonymous.

While in theory the thinner an edge the easier it penetrates, the reality of plants' resistance enters the equation as a limiting factor. An edge with a profile that is too thin will simply crumble. **The process of sharpening can therefore be regarded as a game of finding the compromise between two opposing objectives—easy penetration and durability.**

This is a relevant consideration during both separate, yet interrelated, scythe blade-sharpening steps, which in Chapter 1 I referred to as *beveling* and *honing*.

The bevel of a tool functions as the intermediate portion between its main body and the outermost point of its cutting edge—the initial compromise between the qualities of strength and penetration. Some tools have only one bevel, referred to as a *primary*; others have a *secondary* bevel as well. Either of them can be on one or both sides of an edge. Their exact angles further define the level of the aforementioned compromise.

Most tools acquire the primary bevel during or immediately following the process of manufacture. The secondary bevels are sometimes *intentionally* added by the maker or later by the user. Their purpose is to increase durability as well as reduce honing time by making unnecessary the removal of steel from the entire primary bevel.

To decide on both bevels' specific angles and whether the secondary bevels should be made from either side, both, or neither, is the personal touch of an individual user. For instance, a slight (5 degree) secondary "back bevel" on the *underside* of some broad axes, drawknives or large paring chisels *reduces* their tendency to dig into wood too aggressively, and so makes them more maneuverable. The zone of bevels is, of course, in a continuous process of transformation as any tool is being used and honed.

The scythe blades of Continental Europe, Russia, and the Near East have a clearly defined *primary bevel from the top*. Initially the task of the scythesmith, the beveling is accomplished by a somewhat unusual process of *cold-shaping* referred to as **peening**. This is *the first step in sharpening* taken at regular intervals, by the mower, throughout the continued use of the tool. Its purpose is to recreate the *desired geometry* of the primary bevel. At even more frequent intervals (about 5 minutes) the bevel is further finished by honing, more commonly, in reference to the scythe, termed whetting.

It will be eminently helpful for the user of not only this, but also any other edge tool to understand that honing, by its nature, is a somewhat contradictory process. While the use of the stone is essential to help one achieve the best performance, it gradually wears off the very tip of the edge, eventually necessitating the return to the first sharpening step—the reshaping of the primary bevel.

On the subject of scythe blades, it is a highly pertinent consideration because the honing of this tool is a classic example of *unintentional creation of secondary bevels*, which rather quickly destroys the carefully shaped geometry of the very edge and with it the best performance.

After numerous whettings (from 2 to 6 hours of use) the tip acquires a certain round-ness or convex shape. Although hardly perceptible at a casual look, the change is readily noticed in use. The effect of honing gradually becomes short-lived. At a certain stage the blade may still cut tender plant material relatively well, especially if used with more force.

Eventually the stems of the tougher grasses will not be cut at all, bending over instead; relative to its potential, the tool has lost much of its usefulness.

The alternately shaded areas in Figure 2 b) illustrate the change.

Let us say that ¹⁄₁₆th of an inch was removed from the tip. The bevel from the bottom may end up even a little steeper and more rounded in shape than the top, but so short

Figure 2

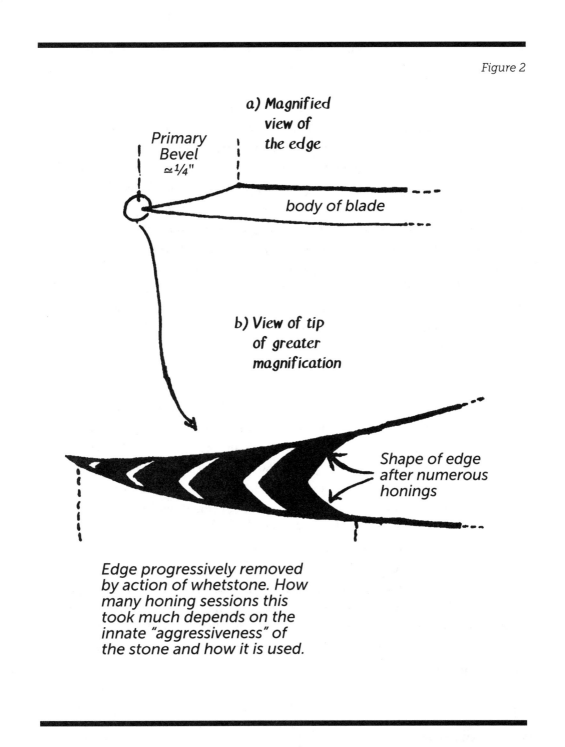

a) Magnified view of the edge

Primary Bevel ≈ ¼"

body of blade

b) View of tip of greater magnification

Shape of edge after numerous honings

Edge progressively removed by action of whetstone. How many honing sessions this took much depends on the innate "aggressiveness" of the stone and how it is used.

that it is hardly noticed. As insignificant as it may seem, it is analogous in function to that of the "back bevel" on a broad ax—both *reduce penetration*. For the cutting of grass with the scythe, however, this is not desirable.

Even though "Austrian style" blades are slightly curved in cross section, **the outermost portion of the edge itself should be maintained flat on the underside**. Peening, if done well, assures this automatically. Furthermore, it can accomplish something which neither grinder nor file will; that is, create a "hollow-ground" effect, which makes subsequent honing faster and the cutting action smoother. It is this which differentiates well-peened edges of scythe blades from those which are merely ground. *The desired geometry of the tip is one of a slightly upward aiming curvature.* Much as the fibers of wood are penetrated easier by an ax or chisel on a diagonal, *less force is required to sever the stems of grasses if the edge approaches them, not straight across, but on an angle and in the direction away from the point of attachment.* A scythe blade with a flat primary bevel will certainly cut grass, if it is well honed. However, an edge as discussed above will cut with more ease. . . . This should clarify why the peening of scythe blades is still the best way to bevel them—not because it is traditional but because it works.

THE QUESTION OF "THIN"

The Austrians use an expression "a good scythe blade edge must run over the thumbnail." What is meant is that if the thumbnail is moved carefully, with upward pressure, underneath the edge for an inch or two, one can see the metal deflect slightly and create a running wave. (This effect can also be seen, though less distinctly, if the thumbnail is merely rocked sideways at one point rather than actually moved along. Safety-oriented individuals can use the rim of an anvil or the edge of a natural stone to do this test.)

It is a poignant description of the high standards required by the mower of old and catered to by the skill of the scythesmith. Only a quality tool can retain such an edge during extended use. However, I do not recommend an edge this thin for many situations. One must first have an understanding that, as such, it is akin to a chisel or plane blade delicately shaped for the purpose of taking paper-thin shavings off clear wood. The careful woodworker would not run such an edge forcefully into a knot—or if he did, he would pay for it with a damaged tool. Can we learn to use our scythe blades as a good craftsperson uses chisels?

A sensible mower in fields clear of rocks would certainly appreciate the above-discussed edge.

The "knots" that threaten scythe blades are either innately very tough plants or, more often, those that early in the season would be cut easily, yet at late flowering or seed stage will crumble the "perfect edge." To cut such material (and it would limit the scythe's usefulness if we did not) we must back off from the potentially most penetrating shape to one with a greater degree of durability. Exactly what degree is grasped as we use the tool with awareness and gain "**The Mower's Sense**."

Listed below are examples of *some* of the cases in which I do not want an edge so very thin:

1. A good portion of the general field mowing on our farm in August when some of the large goldenrod species, often growing in clumps of up to a dozen stems, are in full bloom.
2. Orchards containing plum or pear species prone to send up suckers some distance from the main trunk.
3. Very rocky areas.
4. Unfamiliar yards where I might expect to find, with the blade, objects other than plant material.

For these tasks a blade nearly ready for re-beveling may be just fine if honed more frequently. In addition, one can switch to a coarser whetstone.

The best purpose-specific edge, however, is still one with its underside flat, the top bevel shorter and, of course, not so thin overall; in other words, somewhat thicker than one which easily "runs over the thumbnail but shaped correctly." It will respond better to a fine grit stone, cut easier and be more durable at the same time.

I have focused on the theoretical aspect of peening which is not difficult to grasp and apply. The results are readily observable because they are related to physical shape. There are other, somewhat more abstract, reasons for partaking in this centuries old activity, namely the hardening of the edge and the "building up of opposing electromagnetic charges within the blade." Both of these deserve to be covered, albeit in a lengthier text.

VI

The Practice of Shaping the Edge

WENDELL BERRY, IN HIS INSPIRATIONAL ESSAY "A GOOD SCYTHE"*, SIMPLY STATES that peening "takes some effort and some considering." He was referring to the classical "freehand" method, which Tresemer discusses briefly in his text. Chapter V represents my attempt at the considering. The theory is applicable to the peening of scythe blades in general. In practice, the freehand method takes patience and a good aim with the hammer.

Fortunately, for the benefit of those who are either too hurried or without a good teacher, some tinkerers, many decades ago, invented various forms of a device to make the process easier and the considering less complicated. In English it has been referred to as the "hammerer of scythe blades," "scythe blade sharpening device," or "hammering apparatus." I will here refer to it simply as the **jig**. One of the two models that are now available in America is shown in Figure 3-4.

Whether Tresemer did not know of the existence of the jig or did not think it was necessary, I do not know. Perhaps he was extra talented and fast with the hammer. Or, while not making it explicit, he thought that without using auxiliary devices we have a better chance to preserve one of those skills which are now on the endangered list. If so,

---

* *The Gift of Good Land*, North Point Press, 1981

Figure 3-4

Peening Jig

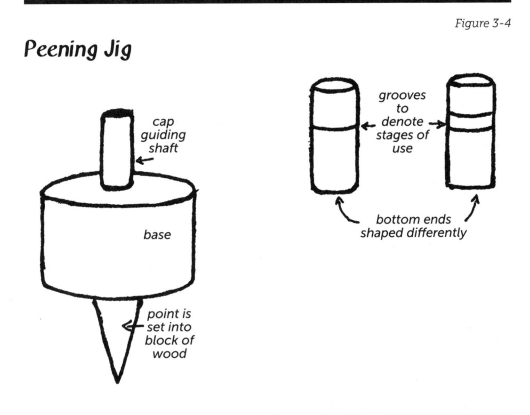

cap
guiding
shaft

base

point is
set into
block of
wood

grooves
to
denote
stages of
use

bottom ends
shaped differently

that nostalgic part of me applauds him, for personally, I too am partial to the traditional method of shaping the edges of scythe blades.

Regrettably, I have come to consider the jig, at least for the average mower of today, an almost indispensable aid. (Many of the blades I have seen in America, "peened thin" by the freehand method, have suffered serious damage as a result of the owners' lack of skill.)

The primary advantage of the jig is that it compensates for inaccuracy with the hammer. One will also be effectively beveling the correct width of the edge without the danger of overreaching. The results will be far more consistent.

The basis of a satisfactory job of peening is the setting up of a steady block at which one can sit comfortably and with good light. The jig can be clamped in a vise, as indeed the instructions on the packaging of some models suggest. I consider this a poor substitute, because few users doing it this way will probably trouble themselves to make an auxiliary set-up to support the protruding portion of the blade while the opposite end is being worked on. With a short blade it would not be as difficult; however, in attempting to steady one of, say, 28" in length while the point is being hammered, the benefit of some means of support will be readily appreciated.

In some parts of Europe it is the custom to peen blades while they are still attached to a snath. Various methods of steadying the snath and blade have been devised. I will not describe them here because I am convinced that more accurate results can be attained if the blade is disconnected from the snath. With the convenience of today's attachment rings it takes only a few seconds. It will also offer a chance to properly re-haft the blade, clean and occasionally oil this part of the snath, which accumulates dirt and moisture, as well as to oil the threads of the setscrews on the ring.

My favorite means of blade support are my legs. They provide the most accurate and instant adjustments of the desired angles of any method I have tried.

To set the jig into the block, it is much easier to pre-drill a slightly undersize hole. The central guiding shaft should *never* be hit unless one of the caps is in place as it could be damaged enough to prevent the smooth movement of the cap. It is also better not to hit the cap hard without the blade inserted; it will retain the desired shape of its lower end much longer if it does not get smashed against the base itself. (I *have* seen children energetically hammering an "empty" jig just for fun.) The blade acts as an intermediate cushion.

Tap the jig into the hole carefully and lightly until about halfway in and somewhat steady. At this point you could insert the blade and begin peening. In time the base of the jig will contact the surface of the block. In the case of the two jig models presently available in America, one complete peening session involves the successive use of two caps with differently shaped bottom ends. The first one will put a defined groove approximately 5/32" (4mm) from the edge. The second will flatten the metal between the groove and the edge and thus also move it a tiny bit forward.

If the blade has been neglected (not damaged per se but left unpeened for too long) it may be necessary to repeat the whole process twice. Done at regular intervals, the first cap may be skipped perhaps every second time or so.

The advertising claims may have the prospective buyer thinking that this "professional sharpening device" will spit out the "perfect edge every time." Relative to the freehand

Figure 5

Blade is touching evenly on both sides of base

Side View

Top View

set into block of wood

method, using the jig is easy, yet success is still not automatic. Deviations from "perfection" can be caused by one or more of several factors. To assure uniformity watch that:

1. The blade is touching evenly on both edges of the base during hammering. (See Figure 5)
2. The very edge itself is well supported while the cap contacts it. (The overall curvature of many models of scythe blades requires that they are tilted slightly forward in order to accomplish this.) (See Figure 6)
3. The rhythm of the hammer strokes is synchronized with the speed at which you move the blade.
4. The force with which you hit is uniform throughout.

The jig requires no specialty hammer. Any common one, of at least 16 oz. in weight, will work. In cases of neglected blades and for the bush models, a 20 oz. will be more effective. I feel it is better to use a bit heavier hammer gently than a light one in too

Figure 6

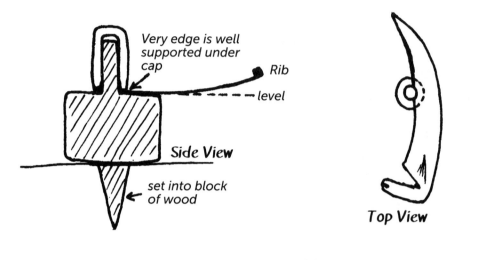

Very edge is well
supported under
cap

Rib

level

Side View

set into block
of wood

Top View

vigorous a manner. **The hammering, even if uniform, can be overdone by hitting too hard**, especially if the primary bevel is relatively new and/or thin (as will be the case with most new blades made in Austria).

Focus on the flattening of the back bevel instead of the thinning per se (which will take place automatically). You may want to make one pass over the blade rather gently to get a feel for the steadying, the rhythm and the extent to which the edge reacts. Then check for the back bevel very carefully. It may still be there in a few spots. Mark these areas from the top with chalk and peen them again, more lightly.

Coming off the jig, the blade's edge, if the job was well done, is suitably shaped (i.e. beveled) but has no keenness. This is due to the fact that during peening it slides against the central guiding shaft, which, of course, dulls it. As a compensation, some of the minute damages too small to be seen with the naked eye are "jointed" or smoothed out. So we are left with an even edge, well beveled, relatively thin, but dull.

Notice the *light reflection* on the edge, which indicates this lack of keenness. Be it a pocket knife or a scythe blade, if the average person's eyes easily detect any reflection while the edge is held toward good light, the tool can be deemed too dull for many of its tasks. Examine the blade edge-on. Flick it slightly sideways back and forth until you are looking at it from an angle where the reflection is greatest. *At no time during normal use should it reflect light along more of its edge than at this stage.* It will now be understood why no more pressure than necessary should be applied with the blade against the jig's guiding shaft.

FINISHING THE PROPERLY BEVELED EDGE

Accordingly, more work with the whetstone will now be required to ready the blade for cutting than during the in-field honing sessions. A synthetic stone will accomplish the first part of this task considerably faster. I suggest that this be done in some comfortable and steady position before mounting the blade. I hone first from the top. With my left hand I hold the blade by the tang, with its point in a block of wood, diagonally across in front of me while I kneel; try other positions if you wish.

For this task of off-the-snath honing of the blade from the top side, use the stone on its narrow edge. (See Figure 7) The majority of scythe stones have a shape that corresponds somewhat to the inner curvature between the slightly upturned edge and the back rib of the blade. Used on their curved edge, they fit within this concavity and thus allow the chance to achieve a lower bevel angle, if this is what the mower desires. As has been explained on Page 115, doing so offers a more penetrating profile but also increases vulnerability. So the true bush blades, for instance, are better off with a less acute secondary bevel and should be honed steeper than the rib-to-edge relationship. (See Figure 9).

If the rib is used as an angle guide, I suggest doing so in such a manner that the stone barely contacts it and the *pressure is focused on the edge.* There is no direct harm in rubbing both rib and edge equally hard (should you find it easier at first) other than needless wear on the stone.

I hone in one direction only, away from myself, from heel to point and on a diagonal, as this allows me a longer stroke. How many strokes this part of the task will take depends on the stone one uses, the pressure applied and how hard the edge was pressed against the shaft of the jig during peening. Continue until you feel a slight "burr" developing from the other side. (For explanation see Page 103)

Those who are new to sharpening tools need to learn to feel for the burr very carefully. It may help if you occasionally rinse the blade or wipe it with a rag. As soon as the

Figure 7

slightest burr is detected anywhere, skip that area and keep honing only the rest. When it can be felt along the whole length, turn the blade upside down and remove the burr.

I prefer to use the stone on its broad face for this task and carefully stroke against (i.e. toward) the edge, as the burr is more quickly removed this way. A smaller portion of the stone will be in use and the stroke will thus be shorter. If you wish for a safer option, do this with the edge away from you. One complete pass consisting of several *overlapping* strokes from heel to point should suffice. Repeat these steps using a natural stone making two or three passes from the top and one or two from underneath. Smooth-cutting edges are achieved through the use of abrasives of a rather gentle nature. Most synthetic whetstones available today do not belong in this category.

All sharpening stones function by cutting grooves in the steel. The larger, harder and/or more angular the abrasive particles (i.e. the grit), the larger the grooves. The bits of steel removed drop off as dust. The inevitable side effect of this process is that some pieces of steel are still attached to, but protruding over, the edge. Most of them will be bent over to varying degrees in the direction *away from the action of the stone*. As the stone is applied from the opposite side some of the burr is bent back again. This game

Figure 8

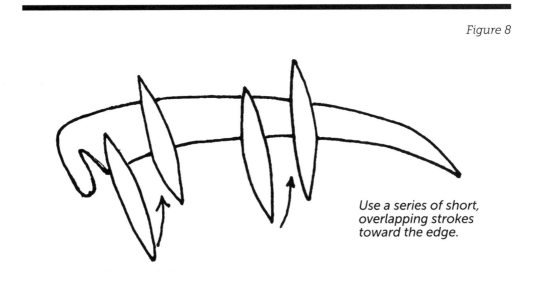

Use a series of short, overlapping strokes toward the edge.

could continue "forever," for even as some burr breaks off, more is created by the grit of the abrasive. (The coarser the stone the more this is true.) An uninitiated person, testing for sharpness of an edge, may feel the individual teeth comprising the burr grabbing the skin of their thumb and the tool would seem deceptively "sharp." In reality, they should be considered a nuisance or at best a temporary condition at a certain stage of a thorough sharpening process.

The purpose of the final honing therefore, has always been to remove the burr as completely as possible. The popular opinion that serrations along an edge are desirable is a half-myth. The objective of creating serrated edges on some cutting tools during the process of manufacture is to reduce the need for the usual frequency of sharpening.

A serrated knife may indeed work very well when new and still cut several years later without having been sharpened. But a distinction must be made. Its relatively sturdy teeth are the result of deliberate *specialty edge shaping by the maker* and are designed to stay there for the lifetime of the tool. The type of teeth we are concerned with when discussing scythe blades are short-lived. At best, some of the smallest and best aligned ones wear off gradually. Most often, a portion of them bend over or break off during the first few swings. When they do, they leave behind a "flat," light-reflecting surface, in other words, *a dull edge at that point*. The coarser the stone used, the wider will be this blunt surface and the less keen the edge. There is no doubt that *edges well finished with finer stones cut easier and smoother*; the burr, being smaller and protruding less, will not bend over or break off as soon.

FREQUENCY OF PEENING

As was explained in Chapter 5, the whetstone starts changing the "ideal" penetrating profile of the edge into a more rounded one from the first whetting on. **The variables affecting the time interval at which re-shaping will be advantageous are therefore all related to how it is honed.**

I nearly always use fine grit natural stones and whet, on average, every 5 minutes. Mowing in a field where only an occasional rock would be found, I peen approximately every 3 to 4 hours. If I used the kind of synthetic whetstones prevalent today, I would halve that time interval.

In many European traditions, 4 to 6 hours was a "day's worth of mowing." During "lunch break" (around 7 or 8 AM) the blades were often touched up with the hammer. Only the men who made a seasonal profession of mowing could be found out in the field during the dry, toilsome hours. Those who could afford to own two blades (both readied the evening before) would switch halfway through the day.

In view of the above, and from my own experience, the 15-minute whetting and 12-hour peening frequencies recommended by Tresemer are exceedingly long. Even if I were a strong man, I would consider it a poor expenditure of energy to coax a blade, maintained on such a schedule, through the grass. Perhaps I am lazy, or simply like to hone and hammer scythe blades. . . . Mowing is, after all, a very personal experience.

The time taken to accomplish the task of peening would not concern me at all. The quality of the results is directly related to patience and attention to details.

Using the jig, it can be done in perhaps 10 minutes (5 minutes per cap). I prefer the freehand method, and after hammering many dozens of blades, still take 15 to 20 minutes for one of average length. It is my very enjoyable centering period of a summer's day.

However, I would find it difficult to cover the intricacies of freehand peening here, so please forgive me for omitting it.

Until more extensive notes are in print, I would like to leave you with one suggestion. Practice the peening (even with the jig) on used "American" blades. They can often be obtained for a few dollars and may save your good blade from abuse. I have yet to come across one that could not be peened; some, in fact, respond better than blades of recent production. An inexperienced person may crack one (which in of itself is a good lesson and a chance to practice edge repair), but not because they are too hard. Rather, they will be found with edges too thick and rounded from both sides. They will crack or chip if hit while not well supported right under the jig's cap or a hammer. Filing the bevel(s) flatter, initially, will help.

VII

Honing in the Field

IN ORDER TO MAINTAIN AN EASY-CUTTING BLADE IN THE FIELD, IT NEEDS TO BE whetted at very frequent intervals. Depending on the innate toughness of the plants and their maturity, as well as the time of day (see Page 148), my average is *after each 5 minutes of mowing*. It takes me approximately 15 to 20 seconds, excluding the wiping of the blade, which should precede each honing session. I rarely make more than one complete "pass" (from beard to point), applying the stone in short, *overlapping* and alternating strokes and using only the action of my wrist. It is fast and has a nice flow to it, though while learning it is easier to concentrate on *one side* at a time.

It should by now be clear that the focus is the honing of the top side (the one with the primary bevel). The main purpose of the stone's use on the underside of the blade is to remove or line up the "burr." Mindful of this, I apply perhaps twice the pressure with the stone against the beveled side, even though an observer might not notice it. This approach delays the inevitable rounding of the edge's underside, i.e. the creating of the short "back bevel."

Figure 9 represents three basic examples of the "stone-to-blade" angles. The considerations can, in fact, become a lot more complicated (much like that of woodworkers' chisel bevel angles) because, apart from understanding the nature of the material being cut, the angles depend on exactly how much "hollow-ground" effect was put on the edge during peening as well as the curve incorporated into the blade's body by the scythesmith.

Figure 9

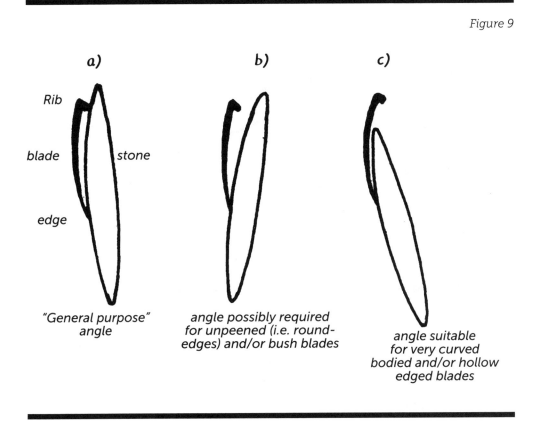

a)

Rib

blade   stone

edge

"General purpose"
angle

b)

angle possibly required
for unpeened (i.e. round-
edges) and/or bush blades

c)

angle suitable
for very curved
bodied and/or hollow
edged blades

Position c) is the one required for the most curved blades and edges. Conversely, on an edge with less curvature, using this position will result in the lowest bevel angle (which translates into better penetration but also less durability.) It necessitates that the tip of the stone does not run along the blade's rib, but rather below it, nearly touching the center of the blade's body. The captions accompanying the stone angle positions in a) and b) will be self-explanatory.

It does not matter which way the scythe is held during whetting, though it is important that the blade be steady and the mower comfortable. The position I prefer is to have the blade across in front of me, pointing to my right, so that I am looking at its underside, as depicted in Figure 10.

Figure 10

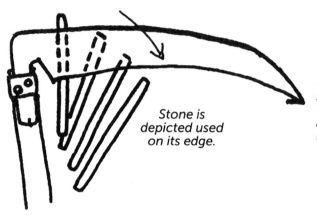

Stone is depicted used on its edge.

Use a series of overlapping strokes from beard to point. They can be alternating or applied from one side only until one pass is completed.

The snath does not need to start, nor during honing remain, in a vertical position. Placing its end on the ground in such a way that the blade's angle allows one to carry out the honing in a comfortable manner will help to maintain the desired stone-to-blade angle and thus improve efficiency.

My somewhat unusual personal technique lies in that I continue the honing directly in front of me while allowing the blade to move gradually to the left by the "creeping" of the fingers of my left hand along the rib of the blade. This works well because the left hand is always very near to where the stone is just then contacting the edge in order to give steadying support where it is needed. There is no periodic pause and re-gripping. Being creative, you will find the way that works best for you.

* * *

Before we leave the topic of easy-cutting scythes, a few reflections may be added here.

There were not many gifts more appreciated by a man of an agrarian culture than a **natural whetstone** of good quality. The mowers of old, however, could tell a good one from a "klumpat." The synthetic versions of 50 or more years ago, though made of much

Figure 11

finer and less aggressive material, were already then referred to as "those which eat scythe blades." The changes in quality have been for the worse. Consequently, I regret to say that most of the synthetic whetstones of recent production worldwide are not suited as the only stone for regular whetting. They are a sort of hybrid between a shaping tool and a honing one, yet they accomplish neither job well.

However, in all the instances where fast metal removal is desired, they will save time; thus they can be appreciated for the following purposes:

1. The initial treatment of a blade peened with the jig. (See "Finishing the Properly Beveled Edge").
2. The removal of the "back bevel" on some brand new blades [see note 5].

3. Edge repairs.
4. Shaping /Beveling (not finishing) of other tools.

Because the classical tapered scythe stones are so easily held in the hand, I use them now for the majority of tasks where abrasives are called for, while the array of bench and other stones I have accumulated over the years is mostly neglected. My carpentry skills are of that mediocre level required on a homestead. The blades I use most frequently (after the snow has covered the grass, of course) are various knives, axes and drawknives, and I do like them "keen" (even if that may not be the same as "perfectly sharp"). Along with farrier tools, sheep shears etc., they are mostly kept in that state by means of scythe stones; several of differing textures are always close at hand.

Having only one scythe stone is, to me, unthinkable; I am totally baffled when I see people quibbling over which one of the $5.00 options they should choose. Men used to risk being shot or thrown in dungeons for smuggling a saddlebag of highly reputed scythe stones over treacherous mountain passes . . .

Luckily a few natural stones are still available. Most of these will outlast synthetic substitutes by several times. From the combined discussion regarding edge quality, it will be obvious that their *slower wear* alone is not the primary reason for their much better value. Furthermore, they can be had without anyone being shot or having to labor a week or two just to afford one, in quality equivalent to the kind that men used to treasure. Such stones consequently received very special care and were kept and carried in equally precious **whetstone holders**.

The first ones, used by the mowers of old, were made of animal horn. With the advent of steel augers wood replaced horn in many regions, probably because it was more plentiful and easily decorated in a personal manner. A well-made wooden holder did not crack, kept the water cool and lasted a lifetime.

Inja Smerdel, curator of the Slovene Ethnographic Museum, wrote a whole book on the subject (as did others before her), and I turn to the words from her text.

> ". . . *Among the various implements, the whetstone holder was a very personal tool with many meanings, and the mower was—regardless of his social status—an individual, a personality with a particular reputation in the rural community. Besides all its*

other meanings, and, of course, besides being simply useful, a whetstone holder was the external token and reflection of a mower's consciousness of his own personality and of his wish to be recognized as such."

How future ethnologists, reflecting on the characteristics of a 21st century American scythe user toting the yellow plastic stone holder, would re-word the above paragraph I dare not guess [see note 6].

VIII

Repairing the Damaged Edge

EVENTUALLY, MOST SCYTHE BLADES WILL REQUIRE MINOR OR MAJOR EDGE repair, and as an opening statement, I would like to emphasize that *damages should be dealt with in tum as they occur and not be allowed to accumulate until such time as "peening proper" takes place.* The shaping of the whole edge and the repairing of a small portion of it are distinctly separate tasks. They can be combined in instances when one damages the edge about the time the whole blade is also due to be peened. In this case, I would complete the repair first, and only then proceed with the actual peening.

For the purpose of repairs, hammers and portable anvil setups were traditionally taken to the field, often shared by a group of mowers. I have also, on occasion, used two rocks to straighten out a simple "dent," a piece of bent but *unbroken* edge. This would be better than ignoring such a damage; one or two rough spots, even if a beginner does not notice the extra wear on the body, will he hard on the whetstone. For repair work alone, you will find the hammer a handier and more accurate tool than the jig. This will get one started, in a small way, on the practice of the freehand method.

If a dent is not very pronounced and especially if bent downward, the jig will work well. When curled considerably, begin by getting it back into place with only the hammer, then finish on the jig with the second cap. More frequently dents are bent upward and are best dealt with "upside down."

Figure 12

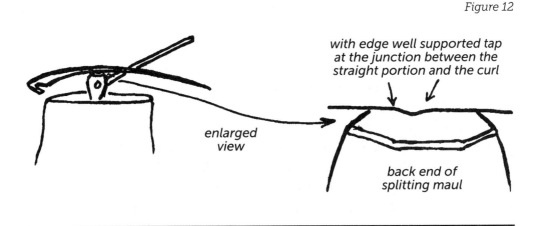

with edge well supported tap
at the junction between the
straight portion and the curl

*enlarged
view*

*back end of
splitting maul*

For this I prefer only the hammer. In the absence of a bona-fide peening anvil, the poll of an ax or a splitting maul driven into a block of wood is a suitable substitute.

The jig will work well in cases where a longer section of the edge is seriously blunted (after being run into a wire fence, for instance). Often it will be somewhat "ragged" as well; file it smooth first, then use the jig.

The more troublesome damages are the perpendicular cracks, as well as "tears," and variously shaped pieces missing from the edge. The latter two are often predisposed by cracks but can also happen independently if the blade hits a sharp-edged metal object (a piece of steel roofing, angle iron etc. forgotten in the grass) or a comer of a partially buried rock. It will likely be a "tear" if the edge was thin; with a thicker, more durable edge, there may be a crescent shaped piece broken out. (The latter, of course, would have required greater force powering the stroke.)

The "cracks" themselves can be caused suddenly, for example when the blade strikes a dry stub of a previously cut sapling, burdock, etc. An edge shaped appropriately for such an encounter should dent. If too thin, it may penetrate partially, but before it severs the stem it will "buckle" or twist momentarily. The extent of the damage will be relative to:

1. The toughness of the plant material as well as
2. The geometry of the primary bevel,

3. The skill of the smith who crafted the blade,
4. The quality of steel it was made of and, of course,
5. The sense of the mower who is guiding the tool.

Admittedly, this is a large dose of relativity and is difficult to consider all at once. Sooner or later most people using the scythe often, or under diverse conditions, will be confronted with this type of damage.

We may end up with several cracks close to one another with just one stroke. The following one, especially if directed at a similar target, will compound the damage. It may extend the cracks, create more of them, remove little pieces between two cracks or along the side of one of them or bend some of them over. See Figure 14 a). An edge so thoroughly "schrupped up" will quit performing as it had been; hopefully *before* the third swing the mower will notice what has happened, stop, and repair the damage.

This may sound dramatic but is no real cause for alarm. A few inches of such an edge will not be much more trouble to repair than a couple of large single cracks.

But before we begin the mending process, let me add that very likely as many, if not more, of the perpendicular cracks are either directly or indirectly caused by the owner's hammer. Overzealous or inaccurate hammering may stress the steel, in one or more spots along the edge, nearly to the point of breaking, until some challenge (perhaps not an obvious one) finishes the job.

The challenge may be more of the same manner of hammering in subsequent sessions. We might cut only young lawn grass in the meantime and not realize what is happening. The cracks may be too tiny at first to be easily noticed by the naked eye, and may not impair the blade's performance seriously. Peen them some more and they will grow longer and more troublesome: corners of the edge next to them bend out and start catching grass. At this point we will take note, but may curse the blade that cracked "on its own," since we are sure we did not run into anything hard.

This is not a hypothetical story, but what some people have actually told me.

Even one session of *careless* peening can take the edge from a state of soundness, through one of stress, to miniscule cracks and then possibly serious ones. It is advisable to acquire the habit of periodically checking the edge under magnification. Even a 3x hand lens will be revealing; 10x is better. *If cracks are detected, hammering in their vicinity*

Figure 13

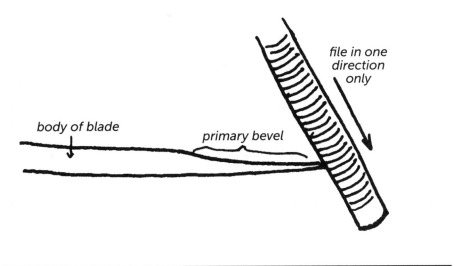

should be avoided. No amount of tapping, however skillful or careful, will again re-connect the intermolecular bonds, i.e. rejoin the two sides of the crack.

To be effectively "repaired," cracks must be removed completely. Here is a job for a file (a half-worn, 6" chainsaw raker file is my favorite).

Before a thorough repair is considered, a compromise approach may be tried out in some cases. A single perpendicular crack, even if quite large (¼" long), can be left intact and watched closely. It may cause no trouble for a long time and gradually wear off in use. Whenever the rest of the blade is peened, avoid hitting the surrounding area—perhaps ³⁄₁₆" to ¼" to either side of it. Instead, shape that part of the edge with a synthetic stone only and keep it a little thicker than the rest.

The kind of diagonal crack we are left with after tapping back the protruding and bent-over piece of a tear is a different story. Once a portion of the edge was stressed this much (by the extreme bending), it will lose some of its "stiffness." The corner may continue to bend out again in use and catch grass, eventually breaking off. Until then it will be a nuisance.

Figure 14

a) *Various examples of original damages: enlarged 2x*

primary bevel

¹⁄₁₆" ¹⁄₈" ³⁄₁₆"

single crack

2 cracks close
to each other with
piece between
broken out

a tear with piece
bent over and
still attached

**b) *the same after <u>inadequate</u> filing: i.e. not filed <u>wide</u> enough; ii) and iii)
also not filed to <u>ends</u> of cracks***

i) ii) iii)

I used to be more concerned with saving every bit of steel and would not file off enough as in Figure 14 b). But in trying to shape the area so it becomes a useful piece of edge again, I would sometimes actually lengthen the crack. This often happens as a result of hammering in close proximity to or certainly directly on a crack. So that the bevel can be well reshaped, the cracks must be filed off completely. However, this must not be attempted over too narrow an area, but rather gradually blended with the existing sound edge.

Figure 15 represents a thorough repair in progress. Needless to say, the taps must be numerous, gentle, overlapping and some will need to reach further up the primary zone than is advisable for "peening proper" (in serious repairs, up to ¼"). There is no need to bring the edge back so it is exactly even.

Figure 15

a)

primary bevel

size of original missing piece

removed by filing

b)

Start placing taps further away from edge
(in case of large cracks, up to the end of the
primary bevel zone, approx. ¼")

c)

Avoid placing taps angled sideways in an
attempt to prevent the corners from protruding.
This will likely cause the steel to "buckle."

Avoid hitting too close to the edge <u>until</u> you
have partially moved more material forward.

d)

Corners <u>will</u> protrude.

← file off →

Partway through the repair file them back a little

e)

Finished repair.

IX

Principles of Movement

DAVID TRESEMER WAS CORRECT WHEN HE SAID, ". . . THAT FOR EVERY 'RULE' OR tradition I have for scythe design and technique, somebody someplace else did it a little differently and got the grass cut."

The commonality of all mowing styles is that the blade moves parallel to the ground and in various degrees of a curve, rather than in a straight line, to assure that the forage is being sliced instead of "chopped." More comprehensive guidelines to the use of the scythe, which I find efficient, easy on the body and pleasant overall, can be condensed to three "rules."

1. Do not carry the weight of the blade, especially during what I have referred to as "field mowing." On an obstacle-free terrain, the only reason to lift the blade is to hone it. Children usually grasp this quickly; many adults, on the other hand, seem to like fighting with gravity and initially need frequent reminding to merely slide the blade forward and back in continuous contact with the surface.
2. Attempt to cut equally to the right and left of your body's center.

There will be exceptions to the above two rules when trimming obstacle-strewn terrain. The following rule, however, should hardly ever be broken.

3. Direct the blade so as to leave all cut grass on the clean stubble of the previous strokes.

To make the practice even more efficient, I would like to add a fourth, not exactly a "rule," but a suggestion. The distinguishing feature of the mowing style I advocate is that it takes advantage of the velocity created by shifting one's weight from one foot to the other. It is most pronounced during "field mowing," yet I still rock sideways (even if only slightly) while performing many short trimming strokes. This is not characteristic of any traditional scything styles I have witnessed thus will be readily appreciated by readers familiar with the principles of movement (and how a significant portion of force is generated) in practically all Oriental martial arts.

Applying those principles to mowing, it deserves to be emphasized that, to take full advantage of the momentum created by shifting the body, cutting less than an eight-foot wide swath is not desirable. It would tire me more if I were to cut a given extended area by adhering to the more traditional width of swath, which is 5 to 7 feet, depending on the region.

Apart from being unable to use much of the innate strength of the legs, I could not synchronize the motions of my body with the rhythm of deep breathing. It is best to exhale fully (through the mouth) while performing the slicing, and inhale (through the nose) on the return movement. (The short trimming stroke does not afford this, though other rhythms are still worth seeking.)

Using the "weight-shifting technique," our 8-year-old daughter, Ashley, can cut a clean eight-foot wide swath through a dense stand. Admittedly, that is an exaggerated width for someone 4'3" tall. The fact is, she enjoys it more than mowing in tangled grass or on a terrain which would restrict this wide movement. The usual width for 11-year-old Fairlight is about 7 feet, and when they mow together I try to have Ashley stick to that also. Often, glancing over my shoulder, I see her gradually stretching it again . . .

Figure 16 is an actual outline of Ashley traced from a photograph (no "stretching" of the dimensions). She is obviously lower to the ground than most of us would find comfortable, but the drawing might explain how someone under 50 pounds in weight can pick up the needed velocity and compensate for lack of strength in the arms.

I do not stand nearly as low or wide relative to my size, though my stance is lower (in the field) than most people seem to have tried.

Nine to 9½ feet is my favored width of stroke, and, on a slight slope with the elevation to my right, I can cut a ten-foot swath. For each person there is, no doubt, an

Figure 16

grass
is cut
all the
way to
here

optimal width that fully utilizes the potential of the weight-shifting principle and yet does not strain any single body part.

Initially, those who are unaccustomed to movements that require this sideways flexibility and the action of the knees may feel that I am overstating the measurements above. Some individuals have actually told me that their legs will not do that. However, without causing yourself discomfort, I suggest that you trust in the innate potential of your body to loosen up. Often I have seen very empowering advancements made in only one focused session.

One of the "secrets" is simply in learning to keep some bounce in the body, especially the knees. To me, all the drawings in *The Scythe Book* of the mower in action portray him rather stiff legged. The only exception is the old painting on Page 34, which I would have put on the front page. My field-mowing stance is even wider, with knees more flexed and the toes pointing in the direction of advance. The distance between the feet indicated in Figure 17 is what I find comfortable; it is related to one's body proportion, leg muscle condition and height.

Figure 17

uncut grass

outline of next stroke

"line"

3' 3' 3½'

cut all the way to the line
under the windrow

slight "spill" of grass over "line"
(edge of field or
previous swath)

windrow

Swath width
9½'

Figure 17 illustrates how Rules 2 and 3 are applied.

As seen, the windrow of grass accumulates within the cut swath for the most part. Only a small portion (varying with how heavy and/or tall the stand is) will "spill over" beyond the line.

If I were cutting through the *middle* of the field, the standing grass to my left would "catch the spill," yet should not be pushed over by it. So, if upon finishing one such pass, I turned left and reversed my direction, I could cut a swath that would lay the new windrow against the first one. The stubble under this rather wide "double windrow" should be free of uncut stems.

The forward movement is as follows: As the cut proceeds, the weight gradually shifts to the left foot. With the slicing portion of the stroke completed and the blade to the extreme left, the right foot, relieved of practically all weight, is free to move forward. The foot is not lifted; it merely slides ahead about the same distance that the blade advances into the grass with each slice. As the weight shifts back onto the right foot during the return stroke, the left one slides forward. It is a slow, continuous creeping forth with no up-and-down movement of the whole body and no real pause or distinct "stepping ahead."

The momentum created by the shifting of weight from the left foot to the right, *alone*, begins to bring the tool back to the starting position. The shoulders as well as the hands' hold on the grips can relax. Only while cutting certain tangled plants will a slight lifting of the blade's heel at the start of the return movement be required, and thus, occasionally, Rule 1 will not be observed.

I consider the technique described above to be the "Basic Form," and believe that a better mower is "made" sooner by practicing it than by deliberately cutting a narrower swath in an open field. The length of one's blade, as mentioned in Chapter 4, will affect the swath's width very little. I can make a 9-foot stroke as easily with a 20-inch blade as with a 30-inch one. Once I approach my "personal limit" (determined by the manner of movement as well as the length and design of the snath), the longer blade will allow me to reach several inches further to the left (but not to the right), and so cut a swath 9½ feet wide.

However, the short blade's limitation is the amount that can be taken off the face of the crescent. This is further influenced by the blade's specific pattern. (The "wide-bearded" models, for instance, are designed for more "depth.") The limit of the "average" 22" to 24" blade in a wide movement may be 4 to 5 inches. A novice may start, therefore, by taking only a 2- to 3-inch depth, increasing it gradually as no uncut strips are left in the stubble.

I would agree with Tresemer's statement "It is no disgrace to be cutting a swath two feet wide" (Page 35 of *The Scythe Book*) *if* he had continued with something like "where space or grass conditions are limiting factors, and *if* the mower stands in the center of that two foot swath." However, I strongly disagree with his advice that "Inexperienced mowers would be wise to take smaller bites by walking a bit to the left, even walking in the cut grass of the previous row." (*Ibid.*) Even if it were physically possible, [see note 7] the mower would find himself working predominantly the right side of the body. As such, it is a fundamentally poor approach through which undesirable habits will be acquired.

The stance and technique depicted in Figure 15 (Page 30 of *The Scythe Book*) is somewhat better, but would still overexert the right arm and shoulder. It is also a waste of energy to throw the cut grass so far to the left. In doing so, only a portion of the potential accomplishment for any given time period will be realized.

TRIMMING TECHNIQUES

Once you experience, and are comfortable with, the dynamics of the Basic Form, you will be able to adapt it to situations of great diversity. It is difficult to explain on paper how creative the cutting patterns can be, when mowing very irregular areas strewn with rocks and junk and interspersed with trees. You can even trim those tall and uneven mounds of topsoil made by bulldozers and often left to grow up in "weeds."

As was noted in Section IV (Blade Selection) it is much easier to mow uneven surfaces with shorter blades because they can follow the contours more accurately. The habit of lightly "hugging the ground" most of the time will be appreciated here.

At times, some of the previously stated "rules" will need to be broken. The shape and size of the strokes, as well as how you stand in relation to the cut, may be continually changing. One can follow a few short movements (between narrowly spaced trees, for instance) with a 9-foot stroke as soon as space permits, cramp the pattern and possibly use a short chopping motion, then widen again for a spell. All this can be done in a relaxed manner, provided the blade is sharp and you learn to follow "the path of least resistance."

That path is rarely marked by signposts but is always there for the attentive mower to follow.

Apart from all else I have discussed thus far, the ease of cutting is affected by:

1. The Time of Day

By far the easiest time for a blade to penetrate the stems is just prior to dawn. It has little to do with actual dew, though its presence is often indicative of the plant's *internal moisture content*, which changes over the 24-hour cycle. With a gradual rise in air temperature the cells transpire water and shrink. At sunrise I would hone the blade perhaps twice as often as at daybreak. By the time the dew is completely gone, any wise mower would be resting in the shade . . . his day's mowing done.

2. The Density and Maturity of the Stand

The most challenging to cut, in terms of density, are the extremes:

   a. A very *dense*, frequently mowed lawn or an intensively grazed pasture.
   b. Scant growth with unsupported individual stems.

Regarding maturity, burdock (Arctium sp.) can be noted as an extreme example. *Before* the flowering stage, regardless of height or stem diameter, it can be cut with the lightest grass blade models. To deal with any sizable stand of burdocks *at full maturity*, I would trade even a sturdy bush scythe for a sharp ax or a pair of long-handled loppers. The above principle is true, albeit to a lesser extent, with practically every other plant, though most of them can certainly be cut with grass blades *if the edge is suitably shaped and the mower's technique correct.*

3. The Direction of Approach

Moving predominantly in straight lines or all the way around the field may not be desirable in many instances. The scythe is not limited to such an approach, and *mowing can often be transformed from hard work to pleasure simply by changing directions.*

Here, two factors, considered together, will determine the easiest path: *topography* and the "*lean*" of the vegetation (even if not apparent, it is *usually* there) within a given stand.

My initial material on the topic of the path of least resistance was many pages long. As with anything else I write, it came under the scrutiny of our 18-year-old home-schooled ("self-taught" would be more accurate) son Kai. He is my first and most critical editor. Apart from cutting up my lengthy Slavic-style sentences, he attempts to take out all the sentimental stuff and many philosophical passages. With his strong tendencies to practicality, if he wrote this addendum it would be one-third the length and titled: "No Nonsense Guide to Mowing!"

Faye is predominantly a doer, not a talker, though occasionally she pens down, in a few moments, what I may have struggled with for days. After the three of us spent several evenings debating how to condense my extensive notes on the "Easy Path," she picked up a pencil and wrote the following:

"Creating a path with this tool, awareness becomes the key concept . . . considering the land, judging the lean, reflecting on the plants, examining the edge, musing over angles, noticing the stubble, contemplating movement, meditating on the breath . . ."

So there you have it—a concise recipe. Short, accurate and yet so expressive of the fact that what I am attempting to write about is more than grass cutting. It is the peasant's dance, the morning prayer under the sky . . .

Axioms such as "If you cannot *rest yourself* while mowing, you are not doing it correctly" attest to the understanding that the task can be almost effortless. [See note 8.]

The practice of T'ai-Chi Chuan, with its emphasis on the executing of smooth motions while in a reflective state of mind, is a beautiful—even if idealistic—model.

The concept of not "losing" the expended energy but rather guiding it in returning circular patterns, the manner of breathing and the related imagery I consider all very applicable to the art of hand mowing. Can we, one day, learn to guide the blade's edge through the spaces between the plants' cells? I think it is worth a try . . .

X

The Blade-Fitting Challenge

FROM SOME OF THE STATEMENTS I HAVE MADE IN THIS TEXT, THE READER MAY surmise that my opinion of how scythes are often fitted, adjusted, sharpened and used is rather low.

To express the overall picture to a farmer 100 years ago, I might have stated that the average scythe of today works somewhat like an underfed and poorly shod horse wearing a collar two sizes too large. This is only loosely descriptive because the majority of contemporary mowers, or mowers-to-be, never drove such a horse.

The analogy to a car may be more fitting, since the average person nowadays does know how it feels to sit behind the wheel of a smooth-running one. Using this metaphor, I believe very few scythe outfits run as if "on all cylinders." Most of them need a push or a pull to make it up any serious hills, and, with respect to comfort features, are like driving a vehicle without shock absorbers.

Take this as a comical exaggeration if you wish—it may lighten your heart, which alone can help with the mowing. However, I am serious, and also certain that until some elemental challenges of a fit are addressed and dealt with, for many mowers to maintain their body in a comfortable position while also keeping the ideal edge-to-ground relationship as illustrated in Figure 20 a), will be mutually exclusive goals.

Figure 18

a)

"a" affects the lay of the blade. Disregarding its variations has been most responsible for many unit's innate disharmony.

a

b)

"b" affects the hafting limits. Usually if the one blade model a dealer chooses to sell fits the specific snath, it will fit all mowers unless their style of movement is distinctly different.

b

The concept of "c" I find too complex to tackle here.

c)

c

What has not been generally considered by the scythe sellers, as well as the buyers, is the fact that *no blade model's specific tang position is well suited to all mowers using the same model of snath, even if the latter is "custom-sized" or adjustable.* Some combinations work better than others more often or for more people.

WHY IS IT SO COMPLICATED?

Apart from all other characteristics that define each blade as a *distinct model*, its tang is angled simultaneously in three specific planes.

The scythesmiths do not invent these angles. They have always followed the exact specifications of their customers. Each respective model was "dreamed up" by someone who had reasons of his own for wanting it one way or another. Many hundreds of models, by the way, have been conceived and made. The Styria Company alone, for example, has patents for over one thousand of them. The specifications for each are in their records and so could all be reproduced. There is no need to do this for, even though the demand for diversity here (as with anything else) is steadily dying, they still manufacture over 200 models; plenty to choose from. However, we—and I refer here to both the dealers and the mowers—need to learn what to ask for and *why.*

Whatever the benefits of the "Information Age" have been in some areas of human endeavor, it has obliterated much wisdom, skill, and service elsewhere. The scythe has been a prime example of the latter, and it is a worldwide condition. As small regional dealers gradually disappeared, many traditional blade and snath models have been replaced by others. Adaptation of the specifications and measuring methods have not kept pace with these changes. Some of the former fine-tuning considerations have been forgotten and many casual mowers do not have their units well adjusted.

May it never come to this, but if the present trends toward less diversity—in general—continue until the world's scythe users are offered only one model of snath and blade, and if we all mowed in the same style, then very simple formulae could suffice. At this point we are caught in mid-stream, with many models of blades and snaths circulating in a haphazard manner, sold in poorly matched pairs, and without the knowledge required to make the best use of them.

The curved metal snath—an easy one to manufacture commercially—is the biggest seller in Continental Europe and is distributed in regions which still use some of their traditional blade patterns. This snath does not fit all of these well, but many people buying them today do not realize this. The decisions as to what should be sold often come from "above"—the large wholesalers who know little or nothing about mowing. What determines their choices, above all other concerns, is what product can be had at the lowest price. [See note 9.]

The exceptions to this are very few; the Sahli Company of Switzerland is one of them. Here, *each single blade is individually matched with a snath from over 20 different wooden models* sold in the respective regions where they are still preferred.

This is exemplary service, rare in today's world. It continues in those areas of supply where obvious benefit, in terms of greater immediate profit, is to be gained by it—i.e. the customers are willing to pay for it. Those who would buy a scythe are assumed, by the world of commerce, to belong to one of two categories:

1. They *need* the tool, which must mean they are members of an "unprogressive" nation and so too poor to pay a high price.
2. They are casual users in wealthy countries who cannot tell the difference between high quality and its substitute, or proper and improper fit. Thus, for this second group, it does not pay to carry the expensive line or many options either.

What makes Sahli special is that they would probably get the same price per snath/blade unit even if they did not pay the salary of a man who does nothing else but match them. The company's reputation was built on "service first," and they are sticking to it.

Today the knowledge of *how* to fine-tune a scythe is becoming rare. Many of the dealers have continued ordering their models, as before, in one tang angle, not considering the fact that today's customers are a different lot.

The existence of adjustable snaths—a relatively modern invention—partially obscures the issue. Their purpose is to cater to mowers of different heights rather than variations in tang angles. A snath designed *also* with the latter in mind would need, in addition, a hinged joint somewhere near the bottom.

In use, the snath meets the ground surface at various angles. Attempting to keep the edge-to-grass stem relationship at its best, the mower continually compensates for changes in terrain and the blade's distance from the feet (hopefully by altering the stance and lifting or lowering the arm, rather than bending over).

What bears consideration when matching units is that the *average* of this snath-to-ground angle will be higher with taller people. In a trimming stroke close to one's feet, the difference may be as great as one degree per inch of height!

To specifically discuss each combination of snath/blade model sold in America today is beyond the scope of this treatise. Though I am hoping to inspire the individual dealers to give this matter some thought, we can *collectively* begin with a few elemental considerations.

In addition to the correct snath size in relation to a person's height and for its intended purpose, there are two important features which distinguish a well-matched unit from a poor one: the ability to achieve a favorable *"hafting angle"* and the *"lay"* of the blade.

CHECKING THE BLADE'S LAY

Holding the scythe in front of you as if you were mowing, contact the surface with the blade as indicated in Figure 19.

Figure 19

The board and point will be further from the surface than center of blade. Keep the point slightly higher than beard.

Contact the ground with this area of the blade

View in Figure 20 is in this direction

beard

point

The relationship of this portion of edge to surface is indicated in Figure 20

Figure 20

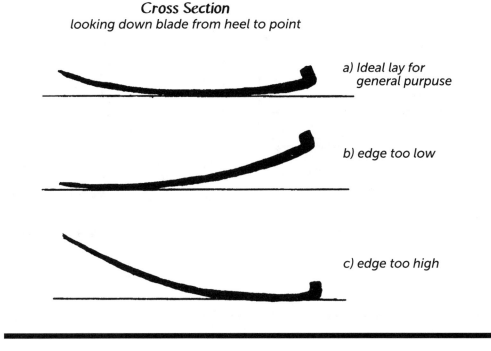

Cross Section
looking down blade from heel to point

a) Ideal lay for
   general purpuse

b) edge too low

c) edge too high

The blade should lay as in Figure 20 a).

A short mower, particularly if using a curved snath with a bush blade (the tang of which is generally steeper), may experience condition b). If adding a wedge, as in Figure 21, is not enough, acquiring a straighter snath or a blade with a less steep tang may be in order.

People of up to 5'8" tall are served most adequately by the present suppliers, though all straight snaths sold until now with grass blades, used by anyone taller than this, probably need the bottom 3" trimmed as in Figure 22.

For many tall mowers this will still not be enough; adding a wedge, to further increase the taper at the snath's lower end, could be the next step. Beyond that, a self-made snath is, from my perspective, the best solution.

Figure 21

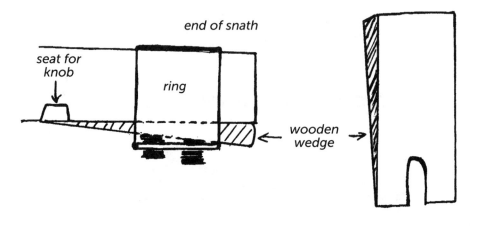

seat for knob

end of snath

ring

wooden wedge

Figure 22

remove in gradual taper to prevent ring from slipping forward as it is being tightened

½"

leave at least ¾" of wood

remove

HAFTING ANGLE

This is the distance relationship between the center of the upper grip A), the blade's beard B), and its point C).

The style of movement, as well as the snath and blade models used, determine exactly what the difference (if any) should be between A–B and A–C. Please note that the guidelines indicated by Figure 5 on Page 11 of *The Scythe Book* will work only if the hafting angle is measured in relation to exactly where the left hand is while mowing. To use the end of the snath as a reference point around which to rotate the unit would only be accurate if the upper grip also crossed that line, i.e. was attached so as to form a "T" at the upper end of the snath.

If you chose this method of hafting, i.e. standing the unit against the side of a building, I suggest an extra touch. Use a doorframe or, if the wall is wooden, drive a nail 4" to 6" from the ground upon which you can rest the center of the grip. In addition, this will prevent the end of the grip from hitting the ground and so distorting the measurement.

If you do not think that 3" one way or the other could make much difference, tape or tack a piece of lath to the end of the snath, rest each position (marked by Xs in Figure 24) in turn upon that nail and see how this affects the hafting angle.

It may be simplest and most accurate to do the initial measuring with a tape and with the unit lying on the ground. Ultimately, all experienced mowers would do the fine-tuning feel and only guess the distance by eye. However, I do urge you to experiment, preferably by practicing the wide stroke on a dense lawn until you find a position in which the cutting is smoothest. It will be a worthwhile accomplishment because, for each specific blade design, in relation to the pattern of cut it is guided through, there is an almost magical adjustment at which the resistance offered by the grass stems will be lowest, while the potentially largest area is cut with each stroke.

Once you think you have found that "best" setting, put a distinct mark behind the tang, just above the ring. It is useful in alerting the mower to the fact that the blade has slipped within the ring. During each whetting, the mark can be glanced at.

At the risk of being considered a perpetual complainer, I will say, in all honesty, that there is hardly a blade I have seen in America that is well hafted. In many cases it is neglect; the blade has moved back, the mower did not notice the change and, during subsequent use, accepted the setting as "normal."

Figure 23

For most of the blade patterns sold in America, the snaths with grips pointing toward the mower, and the style of movement I advocate, the distance from A to C needs to be shorter than A to B by approximately 1½" for 24" blades, 2" for 25" blades, and 2½" for 32" blades,

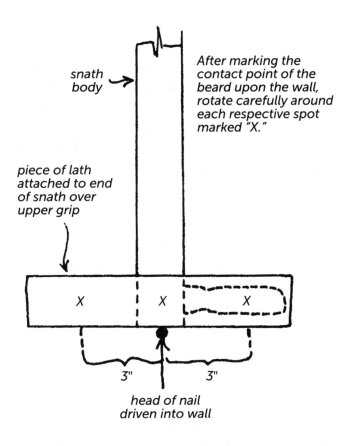

Figure 24

snath
body

After marking the
contact point of the
beard upon the wall,
rotate carefully around
each respective spot
marked "X."

piece of lath
attached to end
of snath over
upper grip

X   X   X

3"   3"

head of nail
driven into wall

The bottom of a wooden snath swells with moisture when used in wet grass and shrinks again while hanging in a dry shed. Unless the screws in the ring are carefully tightened each morning before use, the blade *will* eventually slip. Once the back of the tang is allowed to contact the ring, we have a potentially destructive set-up. Using that contact as a fulcrum, the knob of the tang will act as an effective woodsplitting device the next time the blade is stressed (hitting a dry stub of a sapling, a rock or even a dense clump of grass). Many wooden snaths have been ruined in this manner; the ring will often be damaged at the same time. . . . A hafting angle that is too "open" may be less noticeable or even work fine in scanty growth, tender grass, or if using short strokes. In serious mowing it will "drag" and require more energy from the mower.

With certain blade/snath combinations, it is not easy to achieve the desired hafting angle without some alterations. This is because *many blade models sold in America were designed for snaths that curve to the left and/or those with one or both grips pointing away from the mower.* Their tangs can, however, be ground or filed (from the front side) to make them look more like the tangs of "American" blades. It will help (albeit not in every case) with "narrowing" or closing the hafting angle. (This, by the way, is all that is required to make the "Austrian style" blades fit within the confines of the American ring. Not that I recommend this continent's traditional snath; it is an example of a very un-ergonomic design.)

On the other hand I could tolerate a classical "American" blade. Apart from maintaining its edge with the hammer, I would make a suitable handle for it. . . . which is why I much regret to be omitting a chapter on snathmaking here. Making your own is the ultimate way to solve the blade-fitting challenge. This alternative, if it became commonplace, would eliminate the responsibility of the dealers to provide their customers with an "individually-fitted" unit.

You see, I can walk into, for example, an alder thicket with only a hatchet and return in a short time with a piece of wood from which a snath body can be made to personally suit a 7-foot-tall mower using the flat "American" blade and wishing to mow with a straight back. [See note 10.] This is not an exaggeration. The point I am trying to make is that Nature offers us snaths half-ready, so to speak, in any curves desired. They can be finished in the bush, if need be, with a hand brace and a couple of bits, in addition to that sharp hatchet. (At home I use a shaving horse and a drawknife.) Round holes for the grips are easier to make than square mortises and are perfectly functional.

Figure 25

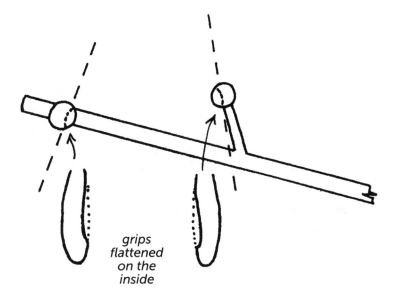

grips
flattened
on the
inside

The combined considerations of a very nice fit, ergonomic design, and possible artistic touches can be involved or, I might add, highly interesting.

A small book could be written on this topic alone. How many people would make use of it, I do not know; I have been told that most present-day mowers will not bother making their own snaths. In any case, I leave you without a bona fide snathmaking chapter for now. One can be added to the next edition if interest justifies it. However, the perceptive reader will find a wealth of subtle visual hints in Figure 1. Here are a couple of additional suggestions:

I would not tolerate round-shaped grips, especially on a trimming snath. They necessitate more "gripping" than is desired for a relaxed experience. Flattening only the "inside" of each grip (as pictured in Figure 25) is a quick and worthwhile improvement to any commercial model. Finish the flattened edges on an angle that feels most comfortable.

Another feature that will help to *balance out the blade* is adding 5-10 ounces to the end of the snath. Slip a pretty brass knob (from an old harness hame) over the tip, or drill a hole and tie your pet rock (not a pebble) somewhere near the upper grip. Experiment with the weight; it is related to snath and blade size, their respective weights and intended purpose.

It took me several seasons to fully appreciate this concept. Now I am very puzzled why this "question of balance" has not been given attention by snath makers and users everywhere.

Discussing this "upside down" idea in theory with various individuals in Europe had little effect; it was too untraditional. But last year, when, for the first time, I traveled during the grass-growing season, I took several home-made snaths along. Every experienced mower in Austria and Switzerland who took one of them, with blade mounted, merely into his hands, nodded in approval. Most of those who took a few swings asked if I could make one for them also. . . which I did.

The shape and placement of the grips were additional features that they all appreciated.

The Appendix outlines some of my attempts to inspire better service on the "home front," as well as my vision for a more promising tomorrow.

Postscript

In spite of my many unflattering remarks, I see the future of the scythe as a bright one. My optimism is periodically renewed by meeting individuals whom I perceive as a symbol of hope.

Eighteen-year-old Hannah Sawyer, from Kingfield, Maine, exemplifies such a person and I offer here a paragraph from her home-school essay:

> *"Scything is simple; it is relaxing too. It is no wonder the people of yesteryear had very little problem with stress. The quiet rhythm melts away tension; swish, cut, swish, cut, swish. . . If I close my eyes it is easy to imagine myself back in time—a bird singing to the sky, the wind whispering to the trees, the splash of a horse pawing the water, the lowing of a cow calling her calf. Would you like to time-travel? I'll teach you how."*

Acknowledgments

Any non-writer who, like myself has attempted to put many elusive concepts and convictions into the fixed form of a book and was left with his immediate family intact, would know that it is the supportive kin who deserve endless gratitude.

Apart from the vast number of people contributing to who I am and what I do, I only know it all started with my mother, Maria and my father, Jozef.

Many have aided my scythe-related quest. Among them, the "amazing woman" Grith Lerche of Denmark and the New Hampshire dairy farmer and lifelong scythe missionary Hans Eccard seem to stand out as wellsprings of certain strength I have, at times, sorely needed.

Moreover, had it not been for the overwhelming hospitality of the extended working collective (and especially Peter Wagner) at the Franz de Paul Schrockenfux scythe factory, where I learned so much, this written form of my partial know-how would not be.

Nor would it be in your hands now without the necessary arrangements made by Alan Hood and the great patience he subsequently manifested.

This manuscript's first versions were all pencil-written on the backs of once-used papers. After my combats with Kai were settled (or nearly so), Faye typed it up on a manual typewriter. Our friend, neighbor, second editor and proofreader Philip Allen, worked long hours to meticulously transfer the text and the diagrams into—well, yes, I sheepishly admit—a "clear digitized form."

Any lack of clarity in composition you may have been troubled by is due to my stubborn insistence on retaining a mode of expression with sometimes uncommon double meanings and "between-the-lines" philosophy. The Slovak brain patterns must have long taproots difficult to dig out.

In spite of the integrity with which I have attempted to communicate, I alone am responsible for possible shortcomings regarding the conceptual and technical correctness of this text.

Appendix

Between 1994 and the time of this writing, I have approached every American scythe importer, including most of the dealers of the "European pattern" blades and accessories, offering to help. (This was free service; I paid for the hours on the phone and spent time writing many pages of letters.)

Some took me for a rare fanatic and treated me accordingly; some seemed to listen politely, yet continued carrying on as before. Still others wanted suggestions (or so they said), but did not follow them or changed their level of service comparatively little.

Unanimously they were of the opinion that:

1. Customers would be confused by too many options.
2. The interest in the tool does not justify the cost of printing more detailed catalog descriptions, or the stocking of a diverse inventory.
3. Their selection, incomplete though it may be, will suffice for the "beginner."

Perhaps I am a naive optimist, partial to the scythe, who believes this tool deserves to be treated on par with those articles where the above-average quality—in terms of service as well as that of the item sold—is in demand and therefore provided.

The biggest "un-success," to date, I consider to be my experience with a Canadian company which, within the last 20 years, has come to represent exemplary tool quality and service in the realms of woodworking and gardening.

Although I did manage to "talk them into" adding the scythe to their offerings, I now wish I had not even tried. When their next catalog arrived, I was shocked at the selection: one blade, aluminum snath, stone-holder and one synthetic stone. This may be expected of an impersonal chain store, but not, (by me, at least) from a company that offers, to the world of woodworkers, approximately 250 different chisels and gouges and a vast array of means to sharpen them.

You see, I had, in several previous letters, urged them to sell shorter blades than 28", and suggested that the dependable Amish woodworkers in Ontario could make snaths of Canadian wood and of a better design than any commercially available. I volunteered to write—free of charge—a short instructional booklet to be included with each complete scythe sale; they did not take me up on the offer and only later did it dawn on me as to why not.

I would have stated clearly—as I did to the company—that these blades will perform poorly if the edge is not periodically "peened." Exasperated, though not yet ready to give up, I called the president on the phone. He told me that, at most, four people in this whole country use the scythe seriously, and, besides, one could not teach the average person today how to cold-shape the edge of a tool, even with the aid of a "jig," (the function of which I had previously described to him in detail by means of diagrams in my lengthy letters.) That really floored me. This man's low regard for his fellow Canadians was one thing; the other is that he cannot claim ignorance on the issue of sharpening. As one of the contemporary "keen-edge" gurus, he wrote an excellent book on the subject. Luckily, he did not discuss the scythe; I wonder how he would have explained, or justified, the use of the whetstone his company sells to sharpen it. Their supplier of scythes and accessories offers five natural and five synthetic stones. They opted for the very *cheapest* of the latter group.

For now I have run out of the energy required to break through established attitudes and business patterns. Instead, I am focusing on an alternative approach—the planting of seeds at ground level with the vision of establishing a cooperative network in America and in Europe. It will consist of those who, instead of just selling the scythe, would like to *learn how to use it well* and, by means of workshops, *pass the skill on to others.* A book is no substitute for a good teacher with the tool in hand, but it is a start. Tresemer's *The Scythe Book* is very inspiring (without being threatening) and fulfills a certain important role. My portion may ruffle a few feathers but also offers a more complete instructional text.

I am presently expanding it further, perhaps beyond the level a "casual mower" will appreciate (it will also include chapters on freehand peening and snathmaking). In addition, an instructional video is in progress to fill the gap between now and the "time of many teachers."

Those willing to help with this extended project can contact us by writing to: Vido Family, 1636 Kintore Road, Lower Kintore, New Brunswick, E7H 2L4, Canada.

Notes

The Scythe Book

1. The word *scythe* is a corruption of the more easily pronounced Old English *sithe*. Peter Cousins suggests an earlier Dutch word, *zicht*, as an origin. These three share the same Indo-European root, *sek-*, as "saw," "sword," "sickle," as well as "(plow) share," "schism," "science," and so on. All the derived words have the sense of splitting one thing from another. The modern German word for scythe is *die Sense*, and relates to the IndoEuropean root, *sent-*, meaning "to head for or go": the scythe is a cutting tool with which you go from one place to another. In some areas, scythe is pronounced like "sigh"; there is no etymological relationship save for the similar sound of blade cutting grass and strong exhalation. The French *faux* and the Italian *falce* derive from the Latin *falx foenaria*, without an earlier root.

2. "Mild" steel has less than the 0.2% minimum carbon content of hard steel and is less strong, less brittle, less sharpenable, and more flexible. Geraint Jenkin's *The Craft Industries* describes in great detail the history of blade manufacture in the British Isles.

3. In Marc Chagall's painting of 1911 of his childhood in Russia, *I and the Village* (Museum of Modern Art, New York), the top grip is held by the mower's right hand, the snath rests horizontally on the right shoulder, and the blade points down behind the back. In Winslow Homer's painting of 1869, *Man with Scythe* (Cooper-Hewitt Museum, Washington, DC), the lower grip rests in the crook of the mower's right elbow, the right hand holds the snath just below the upper grip, and the blade is behind and pointing to the right a foot above the ground.

4. There is no mention of left-handed models in the writings on the scythe (with the one exception of a passing reference in the 1921 *Larousse Agricole*). Nevertheless, in some prominent illustrations scythes appear left-handed. An example is Brueghel's "The Haymakers" (1565, Narodni Museum, Prague). A Norman Rockwell illustration of *Poor Richard's Almanac* first published in 1964 seems to show a left-handed scythe, and has provoked many chuckles from old-timers (Country Beautiful, *The Most Amazing American: Benjamin Franklin*, page 38). These apparently left-handed scythes may, however, have been due to the artist's wish to render a harmonious design—many illustrators show faulty scythe technique, though in pretty pictures.

5. Resting the snath in the crook of the left elbow greatly restricts the range possible with offset handles held by the hands. Nonetheless, this method is shown in the scene for July of eleventh-century calendar from Britain (E. M. Jope, "Agricultural Implements") and surprisingly in modern times in rural Finland (Christopher Williams, *Craftsmen of Necessity*). Modern travelers have photographed scythes with only one lower nib at use in Greece and Turkey. H. J. Hopfen (*Farm Implements*) has presented the six

basic regional types of snath, unfortunately without any discussion of the different ways in which they might be used. Steensberg's *Høstredskaber* shows many different kinds of homemade snaths found in the secluded mountain valleys of Scandinavia.

6. E. R. Tichauer and Howard Gage, "Ergonomic Principles Basic to Hand Tool Design."

7. Any primer on engineering can explain these fascinating realignments at the molecular level. Van Vlack's *Elements of Materials Science* is particularly clear. The "fatigue limit" or "endurance limit" of a metal is the point past which it will not deform under normal stress. Thus the steel snath will slowly deform a small amount, and then stop. The aluminum snath does not stop.

8. By Hand & Foot, Ltd., offers the service of cold-working a badly dented blade with a powerful hand-operated press manufactured in Switzerland for just this purpose.

9. An entry in the register of a London parish dated April 23, 1646, reads: "A child found at Mr. Sawyers in the street on a place to whet knives and was named Edward Sharp." This reference to the use of paving stones for sharpening was noted by Rudolf Hommel in *China at Work*; he also found stationary whetstones widespread in China and Japan in the 1930s.

10. A strickle can be observed in use in William Sydney Mount's 1848 painting, "Farmer Whetting his Scythe," at the Suffolk Museum, Stony Brook, Long Island.

11. From Canto VIII of "Damon the Mower," by Andrew Marvell, in the edition of Marvell's works edited by Margoliouth.

12. Tendons store 2500 joules per kilogram of tendon, compared to 900 for yew wood and 130 for spring steel (J.E. Gordon, *Structures*). In the middle ages, the more efficient catapult design of the Greeks and Romans utilizing animal tendons was apparently forgotten. Instead, a simple heavy counterweight was used to toss a lighter weight into the air. The earlier design stored much more energy in a much lighter material. Thus the image of mowing as a twisting clockwheel depends not only on its pendulum aspect— the lifting of a weight into the air thus creating potential energy for the return movement—but also on the torsion of the dock springs.

13. In Imgard Bartenieff's system (*Body Movement: Coping with the Environment*), the scythe stroke as a whole has the quality of a "Press movement" (strong, direct, and sustained). It is initiated with a momentary "Slash movement" (sudden, strong, and indirect) not a jerky motion, but a flexible launch into the effort, much like the movement of a clockwheel.

14. From a letter by Tommy Thompson in October 1980. Thompson is an Alexander teacher in the Boston area who was trained by the late Frank Pierce Jones, who in turn was trained by F. M. Alexander. Jones's *Body Awareness in Action* is a fine work on the technique. In *The Resurrection of the Body*, Edward Maisel has collected the best of Alexander's own writings, and written a long and helpful introduction to them.

15. Raymond Dart, "The Attainment of Poise," and "Voluntary Musculature in the Human Body."

16. Adrian Bell, *Apple Acre*.

17. Irmgard Bartenieff (*Body Movement*) is articulate about the need for this sort of rhythm:

> *Industrialized society often has a devitalizing effect on organic rhythms. Its pressures reduce the flow or exertion recuperation rhythms necessary to the continuity of vitality in activity. . . . when people confine themselves to actions without recuperations or attempt to fit the actions totally to the machine, the human element on which the next levels of creativity depend is abandoned. (page 75)*

18. R. L. Ardrey, *American Agricultural Implements*; George Fussel, *The Farmer's Tools: 1500–1900*.

19. Jacob Bronowski said in *The Ascent of Man*, "The tool that extends the human hand is also an instrument of vision" (page 118). His statement seems especially true of the scythe, since the words *sight* and *insight* (*sithe* in Old English) have the same root as "scythe," *sek-* (see note 1). Through the scythe I have the opportunity to better know my body, and my world, and how they interrelate.

20. Herbert Koepf, Bo Petterson, and Wolfgang Schaumann, *Bio-Dynamic Agriculture*.

21. This seed mix, as well as each ingredient, is available from Hunter's of Chester, Seedsmen, Chester, England. *Hunter's Guide to Grasses, Clovers, and Weeds* has colored drawings of forty-two plants of the meadow, and is available from By Hand & Foot, Ltd. These particular plants may not be appropriate to all areas; diversity is the aim.

22. In 1952 Newman Turner (*Fertility Pastures and Cover Crops*) set out plots of thirty-five different grasses, clovers, and herbs to see which plants the cows preferred. Several of those which the cows grazed most are included in the permanent pasture seed mix given here and are missing from conventional hay or pasture seed mixes.

23. Jim Worthington, *Natural Poultry-Keeping*.

24. Also pictured in Brueghel's "Haymaking."

25. Clarence Danhof, "Gathering the Grass." Philip Wright, in *Old Farm Implements*, gives the wages for each of these tasks in Essex in 1661. While the actual amounts do not mean much in terms of current prices, they can be compared with each other to give an idea of the relative amount of time it took to do each job. The wage for mowing one acre of grass was 1/10; for raking and cocking one acre of grass 2/-; for reaping, shearing, and binding one acre of wheat 4/-; ditto, rye 4/-; for mowing one acre of peas or vetches 1/9½; for threshing two coombs of wheat 1/-. The monetary units are shillings and pence; at that time twelve pence made a shilling. A *coomb* is four British bushels, or 4.13 American bushels.

26. About cowslip-water, Margoliouth quotes from a herbal of Marvell's day: "Of the juice or water of the flowers of cowslips, divers Gentlewomen know how to cleanse the skin from spots or discoloring therein, as also to take away the wrinckles thereof, and cause the skinne to become smooth and faire."

27. I observed these rakes at use in France. The same design is also pictured in Brueghel's painting, "Haymaking," from the Netherlands in the sixteenth century. Rudolfs Drillis ("Folk Norms and Biomechanics") found the measurements of rakes among Latvians to be personalized: length of handle equal to one's height plus the width of a fist, width of the rake head equal to one cubit, and distance between teeth equal to the width of two fingers.

28. Construction of wooden hayforks and hayrakes is detailed in Drew Langsner's *Country Woodcraft*. The many pictures and drawings show well just how difficult the manufacture of simple tools can be.

29. These carts are available from By Hand & Foot, Ltd.; their design features are thoroughly described in the forthcoming *Handcarts Through History*, also from By Hand & Foot, Ltd.

30. As Rudolf Laban said in *Des Kindes Gymnastik und Tenz* (cited by Bartenieff, *Body Movement*, page 69):

    Work and festivities have their rhythms, their laws, and they must be organized according to the sense of these rhythms. People cry for rhythmic power waves—work and festive—to revive and enliven into meaningfulness their suppressed sense of life.

In the 1930s Laban organized "movement choirs" among groups of industrial workers in the cities of Germany; the movements were based on teamwork in agricultural tasks. The Gurdjielf movements practiced in this country originated with these choirs.

31. From James Thomson, "The Seasons," in O. Henry Warren, *The Good Life*. In *The Song Lore of Ireland*, Mason Redfern gives work songs for plowing, smithing, and spinning, that is, tunes actually used *during* work rather than *about* work sung afterwards. Simple repetitive songs which may have accompanied work are mentioned in Cecil Sharp's *English Folk Songs* (including "Mowing the Barley," as well as "Little Sir Hugh" and "The True Lover's Farewell") and *English Folk Songs from the Southern Appalachians* ("Reap, Boys, Reap"). The scene of planting rice in time with a drummer and a chanter at the end of Kurosawa's film *Seven Samurai* is the best example I know of what I term a work song; viewing this scene makes the power of such songs immediately evident. "Negro Work Songs and Calls" (Library of Congress recording AFS LB) also has this power. In Scotland, Alexander Fenton (*Scottish Country Life*) reports that dances which coordinated threshing teams were not performed for audiences even if these observers were familiar to the workers. The same may have been true with mowing, in which case the songs have died with the mowers who were displaced by the machine. If the current thinking in sociobiology is correct, however, the memory of these movements is still in our blood and can be rediscovered through the work.

32. The "elevenses" or midday break has been well pictured in William Sidney Mount's painting, "Farmer's Nooning" (Suffolk Museum, Stony Brook, Long Island), and Winslow Homer's "Nooning" (Wadsworth Athaneum, Hartford). Brueghel's "Harvesters," reproduced at the beginning of this book, shows some reapers enjoying the noontime break while others work on. The artist clearly wanted to depict different tasks from different parts of the day in the same painting.

33. Adrian Bell, *The Open Air*.

34. Ruth Hertzberg, Beatrice Vaughan, and Janet Greene, *Putting Food By*.

35. From the oral history by Bette Silloway and Jessica Wright, *Maple Sugar Trees and Red Oldsmobiles Revised*.

36. "The Tuft of Flowers," as well as "Mowing," is from Robert Frost's first book, *A Boy's Will*, which was published in 1913. A similar sentiment is expressed in the August 18, 1923, cover of the *Saturday Evening Post* by Norman Rockwell. The cover shows a farmer, with the upper nib of his snath held in the crook of his elbow, rescuing a bird whose nest he had disturbed.

37. Many more weeds and uses can be found in Ben Charles Harris's *Eat the Weeds*.

38. Knowles Shaw (1834–1878) wrote "Bringing in the Sheaves" inspired by Psalm 126 and the evangelist A. D. Fillmore. It is a prominent hymn for the Patrons of Husbandry or Grange, a rural social and political organization which attached rich symbolic value to the tasks and tools of American farmers.

39. Peter Ucko and G. W. Dimbleby, *The Domestication and Exploitation of Plants and Animals*. Michael Jacobsen, in *The Changing American Diet*, documents recent trends in the American diet. Henry Miller's "The Staff of Life" is a wrathful appraisal of American bread and the American diet.

40. Greater detail about the different grains and their culture can be found in *Small-Scale Grain Raising* by Gene Logsdon, *Bio-Dynamic Agriculture* by Koepf et al., and the *Cornell Field Crops Handbook*. *Bio-Dynamic Agriculture* has several unconventional suggestions—such as planting a few chamomile plants in the wheat field to prevent lodging—which are substantiated by experiment. Sources of seed

which cannot be found locally are the Natural Organic Farmers Association (P.O. Box 86, Greensboro Bend, VT 05843), and Johnny's Selected Seeds (Winslow, ME 04901).

41. Peter Cousins has presented the evidence in his thoroughly researched *Hog Plow and Sith*. Good resources on the design of harvesting implements before the modern scythe and cradle are K. D. White's *Agricultural Implements of the Roman World*, E. M. Jope's "Agricultural Implements," and Axel Steensberg's *Ancient Harvesting Implements*. G. E. Fussell ("The Hainault Scythe in England") reports that the mowers usually wore leather leggings for protection.

42. Van Wagenen attributes the establishment of this fact to Professor William Henry Brewer (1828–1910), of the Yale Sheffield Scientific School. The first United States patent for a grain cradle was granted in 1823, although local and state patents were perhaps granted before that time. Indeed, the very first patent granted in this country was for an improved scythe, and was issued by the colony of Massachusetts in 1646 (Victor Clark, *History of Manufactures*, volume I, page 48). However, Joseph Sandford ("Sickle, Scythe and Cradle") claims that the American grain cradle was an improved version of the long-rake scythe from Silesia, introduced to this country by German immigrants in the sixteenth century. Leo Rogin (*The Introduction of Farm Machinery*) cites evidence for a cradle used in Virginia prior to 1760, but does not say which type it was.

43. This design is shown in the several books of Stephens, Partridge, and Cousins, as well as the *Larousse Agricole*.

44. H.J. Hopfen and E. Biesalski illustrate an official of the Food and Agriculture Organization of the United Nations, clean-shaven and dressed in khaki safari shirt and pith helmet, demonstrating the FAO cradle to a bearded, turbaned, and robed Afghan farmer. Still another wooden structure that combines the designs of the FAO and the European large bow is shown in *L'Agriculture dans les Beaux-Arts* by the Musee d'Agriculture in Budapest, Hungary, and in Edit Fel's *Gerate der Atimyer Bauern*. Steensberg's *Høstredskaber* shows several examples from Scandinavia of constructing a cradle by tying a hayfork to both nibs of the scythe.

45. Axel Steensberg's later studies (*New Guinea Gardens*) revealed several more types of ancient harvesting implements. Wood cutting knives, made of bamboo or palm, were cut very thin and sharpened on one end to be hacked at the grass. The knives therefore had no separate haft. After they were used up they were thrown away. Stone knives hafted like axes, ancestors of machetes, were struck against the grass, requiring that a heavy stick of wood be placed behind the stems as a chopping block. These techniques were found in both the New Guinea highlands and the Scandinavian Draved Wood experiments in Neolithic swidden (slash and burn) agriculture.

46. Donald Othmer, "Man versus Materials." According to Joseph K. Campbell, "Power for Field Work," continuous human output would be closer to 1/10 horsepower, but this does not change the point of the comparison. Campbell's formula for periods between four minutes and eight hours is: human horsepower = 0.35 - 0.092 log t (t is time in minutes). Thus the human reaper has between 1,900 and 3,800 times less horsepower available than the 190-horsepower combine, but in an equal amount of time produces more than this fraction of the combine's output. In the South, the cradle was used for over thirty percent of the crop in 1939, perhaps because labor was available and inexpensive; elsewhere in the United States in 1939, the cradle harvested less than ten percent of the crop (M. R. Cooper el al., *Labor Requirements*. and A. P. Brodell, *Machine and Hand Methods*).

47. These figures may have been for a scythe without cradle, which Stephens preferred: "Every species of the cereal grains may be mown with the scythe" (*The Farmer's Guide*, p. 341). His measurements compare the reaping times of three implements, and are given below in the English units of acre, rood, and perch. A rod is 16½ feet; a perch is a square rod; and a rood is 40 perches or 40 square rods. 1 acre = 4 roods = 160 perches. Thus the scythe reaped two acres and three roods or 2.75 acres of wheat.

| | Area Reaped in a Ten-Hour Day | | | | | |
| | Wheat | | | Barley and Oats | | |
	A.	R.	P.	A.	R.	P.
Scythe	2	3	0	4	0	20
Smooth sickle	1	1	18	2	2	10
Toothed sickle	1	0	8	2	0	10

Van Wagenen, Johnston, and the *New England Farmer* of 1834 all claimed four or more acres a day for reaping small grains, without giving data for their claims. The many references cited by Rogin range from two to four acres a day for a cradle scythe.

48. George Fussell, *The Farmer's Tools: 1500–1900*.

49. Made of cast iron parts, the very same planters made today can be found still serviceable after decades on many farms in America and Europe. The Planet Jr. leaves a steady stream of seeds behind, in comparison to most tractor-mounted units and the "Shur Grow Planter" which leave one or several seeds at fixed intervals in the row. The first type is sometimes called a sower, the second a drill.

50. F. H. King, *Farmers of Forty Centuries*.

51. Edwin Betts, *Thomas Jefferson's Farm Book*. Henry Best (*Rural Economy in Yorkshire in 1641*) organized his harvest in yet another way: "Wee allowe one stooker usually to three binders and six Sythes."

52. Peter Cousins estimated three worker-days per acre to reap with a sickle, with two worker-days to bind and stook. He estimated one worker-day per acre to reap with the Flemish scythe, and two worker-days to bind and stook. Rogin estimated ten hours of labor per acre to reap with a cradle scythe, bind, and stook, making Jefferson's calculations a little optimistic.

53. Before baling machines, some farmers bound their hay in the same way as grain. As a symbol of British heraldry, the tied sheaf of grain or "garb" is included in insignia of many different regions and times, with great variation in the material and style of binding. Dorothy Hartley says that binders were cut from long strong grass that grew where oxen dropped or urinated. The ties in Brueghel's "Harvesters" are especially high on the sheaf and the straw particularly long, indicating that the straw (probably rye) was being carefully dried straight for use in roof thatching.

54. Cited in Dorothy Hartley's *Lost Country Life*. The Moabite Ruth was a gleaner, and through her hard work won the respect of the Hebrew patriarch Boaz.

55. H.J. Massingham, *Country Relics*. Not least of these benefits is participation in the rhythm of effort and recuperation (see notes 17 and 30). George Sheldon, "The Passing of the Stall-Fed Ox and the Farm Boy," gives a delightful picture of threshing in the context of farming in the Connecticut River Valley in

the 1830s. Compare Massingham's statement with the passionate analysis of Henry Miller's "The Staff of Life." For example,

> *If they knew what good bread was they would not have such wonderful machines on which they lavish all their time, energy and affection. . . . The machines get the best food, the best attention. Machines are expensive; human lives are cheap.*

56. These studies were reported by the Commissioner of Labor in "Hand and Machine Labor." I have relied on the interpretations of this complicated report in Leo Rogin's *The Introduction of Farm Machinery.*

57. Joseph Needham, *Science and Civilisation in China*, volume I, page 242; George Fussell, *The Farmer's Tools: 1500–1900. Seed Processing and Handling,* edited by Charles Vaughan, shows how these very same processes have been performed by modern machinery.

58. Mary Haferkamp et al., "Studies on Aged Seeds." Lela Barton (*Seed Preservation and Longevity*) rejects the reports of grain recovered from the tombs of Egypt.

59. Bishop Ken, *Hymnotheo*, 1711. Also, see Marvell, in the last line of "Damon the Mower:" "For Death thou art a Mower too." Time, Cronos, and Saturn are all names for the giant ruler of the gods before the age of Zeus. Saturn brought agriculture to the world; his standard was the scythe. Many illustrations from medieval texts in C. G. Jung's *Psychology and Alchemy* give a closer look at Saturn with the scythe. Many drawings of Time show the scythe badly dented, which suggests that Time does not have enough time to peen its blade.

60. William Shakespeare, Sonnet 12, preceded by John Milton, *L'Allegro,* 1631. These quotations are from other countries and other times. I have asked myself why, when thousands of scythe blades are sold each year in the United States, there is so much less symbolic expression about it than in Europe. It is perhaps partly because the first American tool has been the ax—the trees needed to be cleared before the grasses could grow. From one point of view the ax was a scythe: "At each swing they cut down a section of timber as a scythe cuts down a swathe of grain" (Wallace Wadsworth, *Paul Bunyan*), "When Paul Bunyan started lumbering in the West, the fir and redwoods began to fall like grass" (Adrien Stoutenburg, *American Tall Tales*). Thus the scythe did not have in this country as in Europe the hundreds of years of use before mechanical agriculture. Anyone using the scythe, however, can participate immediately in the source of so many hidden symbolic references permeating our cultural background.

61. In *The Golden Bough*, Sir James Fraser has given many variants of these harvest customs. To make a Corn Dolly, any pattern of weaving the pliant straws will do. If you wish to try your hand at the traditional English designs, James Arnold's *Book of Country Crafts* and Germaine Brotherton's *Rush and Leafcraft* are good introductions.

62. From the Christian bishop Hippolytus, cited in Joseph Campbell's *The Masks of God: Primitive Mythology*, page 185. The reference is to the culmination of the initiation rites of Demeter centered at Eleusis. Ceres, from ker-, relates to "cereal," "create," and "crescent;" Demeter, from **mater-**, relates to "mother," "matter," "matrix," and "meter." Each emphasizes an aspect of the whole which was worshipped in their names. For a closer look at the symbolic association to the harvest, with implications for the scythe and its place, begin with Joseph Campbell's book cited above, and *Essays on a Science of Mythology* by K. Kerenyi and C. G. Jung, and Campbell's *The Mysteries*.

63. Findhorn Community, *The Findhorn Garden*. There are now many intentional communities which seek to develop this awareness.

The Scythe Must Dance

Note 1. The work-morale within a socialist state, if the utmost productivity mattered, was nothing to brag about. Imagine the lax atmosphere we would have in North America if we all worked for the Government. . .

In retrospect, however, I am impressed by that system's not succumbing to the notion that machines should replace human bodies whenever possible. Shovels, hoes and, of course, scythes were used along side their modern substitutes whenever it made sense . . . which it often still does. There was no unemployment. Building foundations were dug, and acres of cooperative row crops hoed, by hand. Many animals' bellies were, and still are, kept full with hand-mown hay . . .

Note 2. The Marugg Company's "Instruction Page," written for the 1932 catalogue (which, we are told, "still applies today as is"), shows a man seated by a steady block with the blade well supported on his knees. I do like that picture. However, the combined peening-related text is as follows: ". . . these blades are sharpened by beating or hammering out the cutting edge, instead of grinding. This method produces a wide, keen, somewhat hollow, jagged, cutting edge, at the same time hardening it. Any person of ordinary intelligence can, with very little practice, produce a perfect 'Dangel' by the use of a proper hammer and anvil. Various patterns of Dangel outfits (hammer and anvil) are in use. We show two kinds of hammers and also anvils. In using our 'B' style anvil, the blade is laid flat side down on the anvil and the edge is hammered from the upper side of the blade, When using W style anvil, which has a narrow rounded top, the blade is turned bottom side up and hammered from the bottom as shown in the illustration. To prevent heating of the metal, it is well to dip the hammer in water occasionally. A fresh hammered Dangel edge usually suffices for a full day's mowing."

Perhaps "ordinary intelligence" meant something else in those days . . . Indeed, all other short pamphlets enclosed with the scythe by other companies in other languages that I have found to date are similarly lacking in usefulness. Most of them stress that "the edges must under no circumstances be sharpened by grinding" and that the peening needs to be "executed very carefully" . . . but exactly *how* is left to the mower's intelligence to figure out . . .

Note 3. Formerly titled *Keeping Warm With an Ax*, this is a rare one among the modern how-to books. The author is obviously one of those old-timers who *understands* the subject. Moreover, he treats it holistically and can write with clarity as well as humor. The well-explained concept of "frontal zone" alone, no doubt, saved some of my own blood when, several years ago, I made that break from chainsaw dependency to a far more pleasurable time in the woods. I highly recommend this book, even if you do like your chainsaw! Cook uses one, as well; and he writes from both sides of the fence.

Note 4. I am not implying that a neglected ax can be *well-reconditioned* in a few minutes. However, by using a fine-cut file and three scythe stones of progressively finer grit, I could quickly finish the last of the edge of nearly any old ax so that it would readily shave the hairs off my arm. Even the 60 to 70 degree "combined angle" of a splitting maul will do that. The last stone in succession would be a natural one from Slovakia, which is capable of leaving behind a polished, "razor-sharp" finish. A few strokes on a leather strop would be a nice final touch, though not necessary.

I make an exception for those $4.99 "Made in China" hatchets or similar "economy" axes. This I say *not* to belittle the potential skill of Chinese, Korean etc. people to make products of good quality, but rather the Western World's blind inclination to buy cheap items and so support the spread of Wal-Mart-like empires.

Note 5. There was a time when some scythemakers (Styria was one of them) *did* finish the edges to perfection (by the way, women were employed to hone each one individually prior to sale) and the label "Mähfer-tig" (ready to mow) was born. About 20 to 30 years ago this became cost-prohibitive, but the label remained. It is not put on all blades for all dealers, but has, in my opinion, caused considerable disservice (even though it could be said that "ready-to-mow" is relative).

The experienced peasants will simply ignore it. They grew up hearing the old men tell them "Only after the third (or even fifth) peening is a new blade actually nice to mow with." The modern scythe owner, used to being catered to in so many other ways, falls for the label—or the seller's advertising promise . . .

Although the products we receive from Austria today are pre-peened far better than is (or was) the case with scythe blades world wide, there is not a handful of them really "ready-to-mow" in any single lot. When they come from under the hammer, they would be ready for the whetstone. But the last cosmetic touch (to make the edge very even) is a quick, light pass on a grinding wheel. It is this which "de-readies" them again. Most will end up with a *tiny secondary bevel* either from the top or bottom, often both, which is not absolutely uniform. (It is usually most pronounced near the point.) Until these little bevels are removed, the edge will not respond well to a stone, especially a fine-grit and/or a natural one. Thus a mower's first experience will, all too often, not be as encouraging as it could.

Note 6. Whetstone holders of steel have long been available. They could still be sold by retailers who believed that their customers might appreciate a choice. Unfortunately, the large international distributors now offer the plastic version at less than one third the price of the metal one.

But what is the lifespan of a "bargain"? A galvanized holder once carried by my grandfather, slightly dented yet watertight, continues to be useful. Will many of the plastic ones on the shelves today not lurk within the buried garbage a testimony to our ways, 80 years hence?

In Switzerland, though, even more "costly" wooden whetstone holders can readily be purchased. I also, found a small company to make them for us out of pure copper. While not quite a substitute for a personally carved wooden one, it is a start. Few cows are still privileged to live out their years with horns intact. A local slaughter house may be a place for a treasure hunt.

Note 7. Considering this statement carefully, one may see that what Tresemer probably meant was ". . . even walking in the [stubble] of the previous row." The grass of that row, once cut, would end up so far to the left (using the technique in Figure 17) that a mower walking in it could not reach into the uncut grass.

Note 8. It is probably not as common an analogy as some others, because I heard it only from Lambert Schoiswohl, the man to whom Faye's poem is dedicated. It was his father's friend, already an old man when Lambert was a boy, who emphasized this "T'ai-Chi like" approach to what, back then, was a task of necessity for practically every countryside dweller.

Note 9. When we buy an "economy line" product for which a high quality equivalent is available, we rarely consider that, by this, we endorse the notion that "doing something fast is more important than doing it well," or that "hand labor is too costly to be affordable." In many other ways we have all contributed to an enormous loss of human skill. Some forms of craftsmanship required highly intensive training of very young hands by a master and only a small portion of dedicated apprentices reached that zenith of competence. Once the masters are buried, even the world's billionaires (should they desire that quality again) cannot resurrect them . . .

The making of the classical Austrian scythe blade was one of those skills. As hand labor progressively became "too costly," quality had to be compromised. Blades from countries that never reached the highest level of scythesmithing (or gave it up sooner) have, especially since World War II, competed on the international market.

Today the Chinese-made scythe blade, from my perspective, poses a real threat to the survival of whatever excellence remains. (The term "excellence" must now be used with relativity in this "technologically advanced" period of history). When I first learned about its recent invasion of the European scythe market, I wrote to a large German wholesaler (a former maker of high-quality scythe accessories) asking for a sample. The representative was sympathetic when I explained my reasons for comparative trials, but also frank: he suggested that I would be wasting my time, since the quality of those blades was "extremely low."

Contemporary business ethics embrace many elements of unfairness. The Chinese blade is allowed to be displayed in retail stores bearing its "ALPENKRONE" name (rather suggestive of that mountain region's product) with no country of origin revealed on it. It has two additional labels (in the language spoken where it is sold) "Ready-to-Mow" and "Quality Tested." The latter "Golden Label" was once a highly reputed one, truly earned by excellent workmanship. Today the sticker can, apparently, be bought . . .

That "mowing readiness" (of a totally different class than what the Austrian makers refer to when they denote this on their products) is another blatant lie; yet, by using a classical label, the wool is pulled over the eyes of many prospective buyers—especially when the price is less than one third of an Austrian-made blade. An experienced mower knows, simply by holding the two, that the "economy" blade is not a bargain.

However, many rural lot owners are of another generation who rarely use the scythe and know little about it; they buy it as a sort of stand-by in case their string trimmer breaks down. This is, after all, that once-so-essential tool which should hang in everyone's garden shed as an old-time relic, if nothing else.

This adds up to a lot of sales, and the merchants probably make a higher percentage of profit on a Chinese blade than on an Austrian one.

There are only two scythe makers in Austria today, which is nearly a hundred times less than approximately a century ago. They are having difficulty finding replacements for their retiring men, because there is not enough incentive for young ones to start the lengthy training.

The most frequent question a manufacturer receives from the world's large distributors is "Could you make it a little cheaper?"

Note 10. This last note will, I hope, read as a happy one. Once certain concepts of "fit"—actually fundamentals of simple physics—are grasped, you can make a snath to fit yourself, even if you desire to mow on stilts. Though I do not claim to "have it all figured out," if someone showed me a blade of any tang position, I could find the material for a snath to fit it in an alder thicket. (In other areas of the country, it may be provided by a different species, though in this locality that is the place where snath materials are of the most plentiful curves and nearly the right diameter.) Alder wood is easy to work and it is strong enough—perhaps not for the most "energetic" mowers, but certainly for the sensible ones.

In the same thicket I would also find a lower grip with the "extension" already attached—a branch coming from the main trunk, preferably at an 80- to 85-degree angle.

I rarely cut a live tree (or even a branch of one) for the purpose of snathmaking, because plenty of material is already cut on the roadsides by government crews or left in slash piles by loggers.

The world is full of gifts!

Bibliography

Allingham, William. *Fifty Modern Poems*. London: Bell & Daldy, 1865.

Ardrey, R. L. *American Agricultural Implements*. New York: Arno Press, 1972 (1894).

Arnold, James. *The Shell Book of Country Crafts*. London: John Baker, 1968.

Bartenieff, Irmgard, with Dori Lewis. *Body Movement: Coping with the Environment*. New York: Gordon and Breach, 1980.

Barton, Lela V. *Seed Preservation and Longevity*. New York: Interscience Publishers, 1961.

Bayerische und Tiroler Sensen-Union. "Einige Aufschltisse über die Sensenherstellung." *Landmaschinen-Markt*, 1950, 29.

Bell, Adrian. *Apple Acre*. London: Brockhampton Press, 1942.

Bell, Adrian. *The Open Air*. London: Faber & Faber, 1936.

Bell, Adrian. *Sunrise to Sunset*. London: John Lane, Bodly Head, 1944.

Berry, Wendell. "A good scythe." *Organic Gardening*, January 1980, 138–141.

Best, Henry. *Rural Economy in Yorkshire in 1641*. London: Surtees Society, 1857.

Betts, Edwin (Ed.). *Thomas Jefferson's Farm Book*. Princeton, NJ: Princeton University Press, 1953.

Blandford, Percy. *Old Farm Tools and Machinery*. Detroit: Gale, 1976.

Brodell, A. P. *Machine and Hand Methods in Crop Production* (Farm Management Report 18). Washington, DC: United States Department of Agriculture, Bureau of Agricultural Economics, 1940.

Bronowski, Jacob. *The Ascent of Man*. Boston, MA: Little Brown, 1973.

Brotherton, Germaine. *Rush and Leafcraft*. Boston, MA: Houghton Mifflin Company, 1977.

Campbell, Joseph. *The Masks of God: Primitive Mythology*. New York: Viking Press, 1959.

Campbell, Joseph (Ed.). *The Mysteries* (Bollingen Series XXX, Volume 2). Princeton, NJ: Princeton University Press, 1955.

Campbell, Joseph K. "Power for field work." In Diana Branch (Ed.) *Tools for Homesteaders, Gardeners, and Small Scale Farmers*. Emmaus, PA: Rodale Press, 1978, 89–92.

Cannon, LeGrand. *Look to the Mountain*. New York: Henry Holt, 1942.

Clark, Victor. *History of Manufactures in the United States*, three volumes. New York: Peter Smith, 1949 (1929).

Collins, E. J. T. *Sickle to Combine*. Reading, Berkshire: Museum of English Rural Life, 1969

Commissioner of Labor. "Hand and machine labor." *Thirteenth Annual Report*, volume 2. Washington, DC: United States Dept. of Labor, 1898.

Cooper, M. R., W. C. Holley, H. W. Hawthorne, and R. S. Washburn. *Labor Requirements for Crops and Livestock* (Farm Management Report 40). Washington, DC: United States Department of Agriculture, Bureau of Agricultural Economics, 1943.

Cornell Field Crop Handbook. Ithaca, NY: New York State College of Agriculture and Life Sciences, 1978.

Country Beautiful. *The Most Amazing American: Benjamin Franklin*. Waukesha, Wisconsin: Country Beautiful, 1973.

Cousins, Peter H. *Hog Plow and Sith*. Dearborn, Michigan: Greenfield Village and Henry Ford Museum, 1973.

Danhof, Clarence. "Gathering the grass." *Agricultural History*, 1956, 30, 169–173.

Dart, Raymond A. "The attainment of poise." *South African Medical Journal*, 1947, 21, 74–91 (also, *Human Potential*, 1968, *I*).

Dart, Raymond A. "Voluntary musculature in the human body: the double-spiral arrangement." *British Journal of Physical Medicine*, 1950, 13, 265–268 (also *Human Potential*, 1968, *I*, 89-98).

Drillis, Rudolfs J. "Folk norms and biomechanics," *Human Factors*, 1963, 5, 427–441.

Duby, Georges. *Rural Economy and Country Life* (Cynthia Postan, trans.). Columbia, SC: University of South Carolina Press, 1968.

Fél, Edit, and Tamás Hofer. *Geräte der Atányer Bauern*. Kopenhagen: Kommisson der Koniglich Danischen Akademie der Wissenschaften, 1974.

Fenton, Alexander. *Scottish Country Life*. Edinburgh: John Donald Publishers, 1976.

Findhorn Community. *The Findhorn Garden*. New York, NY: Harper and Row, 1975.

Fischer, Franz. *Die Blauen Sensen*. Graz-Köln: Verlag Hermann Bohlaus, 1966.

Fraser, Sir James George. *The Golden Bough* (3rd ed.). New York, NY: Macmillan, 1966 (1911).

Frost, Robert. *Complete Poems*. New York: Holt, Rinehart and Winston, 1949.

Fukuoka, Masanobu. *The One-Straw Revolution: An Introduction to Natural Farming*. Emmaus, PA: Rodale Press, 1978.

Fussell, George Edwin. *The Farmers Tools: 1500–1900*. London: Andrew Melrose, 1952.

Fussell, G. E. "The Hainault scythe in England," *Man*, 1960, 60, 105–108.

Gordon, J. E. *Structures*. New York, NY: Plenum, 1978.

Gould, John. "There's safety in scything." *Christian Science Monitor*, October 6, 1978, 20.

Haferkamp, Mary E., Luther Smith, and R. A. Nilan. Studies on aged seeds I: Relation of age of seed to germination and longevity. *Agronomy Journal*, 1953, 434–437.

Harris, Ben Charles. *Eat the Weeds*. Barre, MA: Barre Publishers, 1971.

Herrigel, Eugen. *Zen in the Art of Archery* (R. F. C. Hull, trans.). New York, NY: Pantheon, 1953.

Hertzberg, Ruth, Vaughan, Beatrice, and Greene, Janet. *Putting Food By*. Brattleboro, VT: The Stephen Greene Press, 1973.

Homans, George Caspar. *English Villagers of the Thirteenth Century*. New York, NY: Russell & Russell, 1960.

Hommel, Rudolf P. *China at Work: An Illustrated Record of the Primitive Industries of China's Masses, Whose Life is Toil, and thus an Account of Chinese Civilization*. New York, NY: John Day, 1937.

Hopfen, H. J. *Farm Implements for Arid and Tropical Regions* (rev. ed.) (FAO Agricultural Development Paper No. 91.). Rome: Food and Agriculture Organization of the United Nations, 1969.

Hopfen, H.J., and Biesalski, E. *Small Farm Implements* (FAO Development Paper No. 32). Rome: Food and Agriculture Organization of the United Nations, 1953.

Horne, Thomas Hartwell. *The Complete Grazier* (5th ed.). London: Baldwin & Cradock, 1830.

Hunter, Peter J. P. *Hunter's Guide to Grasses, Clovers, and Weeds*. Chester, England: Hunters of Chester, n.d.

Ingalls, John James. "In praise of blue grass." In *U.S. Department of Grass: The Yearbook of Agriculture 1948*. Washington, DC: Government Printing Office, 1948, 6–8.

Jacobsen, Michael. *The Changing American Diet*. Washington, DC: Center for Science in the Public Interest, 1978.

Jenkins, Geraint. *The Craft Industries*. London: Longman, 1972.

Johnston, James F. W. *Notes on North America: Agricultural, Economical, and Social*, Volume I. Edinburgh, 1851.

Johnston, Robert, Jr. *Growing Garden Seeds*. Albion, Maine: Johnny's Selected Seeds, 1976.

Jones, Frank Pierce. *Body Awareness in Action: A Study of the Alexander Technique*. New York, NY: Schocken Books, 1976.

Jope, E. M. "Agricultural implements." In Singer, Charles, E. J. Holmyard, and A. R. Hall (Eds.) *A History of Technology*, Vol. 2. Oxford: Oxford University Press, 1956, 81–102.

Jung, C. G. *Psychology and Alchemy* (Second edition; R. F. C. Hull, trans.) Princeton, NJ: Princeton University Press, 1968.

Jung, C. G. *Symbols of Transformation* (Second edition; R. F. C. Hull, trans.) Princeton, NJ: Princeton University Press, 1967.

Kerenyi, K., and Jung, C. G. *Essays on a Science of Mythology* (R. F. C. Hull, trans.). Princeton NJ: Princeton University Press, 1971 (1949).

King, F. H. *Farmers of Forty Centuries, or Permanent Agriculture in China, Korea, and Japan*. Emmaus, PA: Rodale Press, (1911).

Koepf, Herbert H., Petterson, Bo D., and Schaumann, Wolfgang. *Bio-Dynamic Agriculture*. Spring Valley, NY: Anthroposophic Press, 1976.

Kushi, Michio. *The Book of Macrobiotics*. Tokyo: Japan Publications, 1977. Langsner, Drew. *Country Woodcraft*. Emmaus, PA: Rodale Press, 1978.

Logsdon, Gene. *Small-Scale Grain Raising*. Emmaus, PA: Rodale Press, 1977.

Lühning, Arnold. *Die Schneidenden Erntegeräte* (Unpublished doctoral dissertation). Göttingen: Georg-August Universität, 1951.

Maisel, Edward. *The Resurrection of the Body: The Writings of F. Matthias Alexander*. New York: Delta, 1969.

Margoliouth, H. M. (Ed.) *The Poems and Letters of Andrew Marvell. Vol. 1: Poems* (3rd ed.). Oxford: Clarendon Press, 1971.

Massingham, H. J. *Country Relics*. Cambridge, England: Cambridge University Press, 1939.

Miller, Henry. "The staff of life." *Selected Prose*, Volume I. London: Macgibbon & Kee, 1965, 331–345.

Mitchell, John. "Scyther's complaint." In Wayne Hanley and John Mitchell (Eds.) *The Energy Book*. Brattleboro, VT: The Stephen Greene Press, 160–175.

Murphy, Michael. *Golf in the Kingdom*. New York, NY: Dell, 1972.

Musee d'Agriculture. *L'Agriculture dans les Beaux-Arts*. Budapest: Musee d'Agriculture, 1959.

MyJonas, G. E. *The Hymn to Demeter and Her Sanctuary at Eleusis*. St. Louis, MO: Washington University Press, 1942.

Needham, Joseph. *Science and Civilisation in China*, Volume I. Cambridge: Cambridge University Press, 1954.

New England Farmer, 1834, 1:33, p. 22.

Oakley, Kenneth P. *Man the Tool-Maker*. Chicago: University of Chicago Press, 1968.

Othmer, Donald F. "Man versus materials." *Transactions of the New York Academy of Sciences*, Series II, 1970, 32, 287–303.

Partridge, Michael. *Farm Tools Through the Ages*. Reading, Berkshire: Osprey Publishing Co., 1973.

Pfeiffer, Ehrenrend. *Weeds and What They Tell*. Wyoming, RI: Bio-Dynamic Farming & Gardening Association, 1974.

Redfern, Mason. *The Song Lore of Ireland*. Ann Arbor, MI: Gryphon Books, 1971.

Roger, Jean-Marie. "Blé," *Encyclopedie Permanente d'Agriculture Biologique*. Available from the International Federation of Organic Agriculture Movements, Coolidge Center for the Advancement of Agriculture, Topsfield, Massachusetts.

Rogin, Leo. *The Introduction of Farm Machinery in its Relation to the Productivity of Labor in the Agriculture of the United States During the Nineteenth Century* (University of California Publications in Economics, volume 9). Berkeley: University of California Press, 1931.

Roper, John S. *Early North Worcestershire Scythesmiths and Scythegrinders—A Study Based on Wills and Probate Inventories, 1541–1647*. Cambridge University: Dudley House, 1967.

Sandford, Joseph E. "Sickle, scythe and cradle," *The Chronicle of the Early American Industries Association*, December, 1938, 2:7, 49, 51-52.

Schrökenfux, Franz. *Geschichte der österreichischen Sensenwerke und deren Besitzer*. Linz-Donau: Herausgegben von Franz John, 1975.

Sharp, Cecil J. *English Folk Song* (4th ed.). Belmont, CA: Wadsworth Publishing Co., 1965.

Sharp, Cecil J. *English Folk Songs from the Southern Appalachians*. London: Oxford University Press, 1966 (1932).

Sheldon, George. "The passing of the stall-fed ox and the farm boy." *Proceedings of the Pocumtuck Valley Memorial Association*, 1898, 111, 5–29.

Silloway, Bette, and Jessica Wright. *Maple Sugar Trees and Red Oldsmobiles Revised*. Chelsea, Vermont: Acorn Press, 1974.

Steensberg, Axel. *Ancient Harvesting Implements: A Study in Archeology and Human Geography*. Copenhagen: Bianco Lunos Bogtrykkeri, 1943.

Steensberg, Axel. Höstredskaber. Lyngby: Nationalmuseets Etnologiske Undersøgelser, 1941.

Steensberg, Axel. *New Guinea Gardens: A Study of Husbandry with Parallels in Prehistoric Europe*. New York, NY: Academic, 1980.

Stephens, H. *The Book of the Farm* (4th ed., revised by James MacDonald). Edinburgh: William Blackwood and Sons, 1891.

Stephens, H. *The Farmer's Guide*. New York: Leonard Scott, 1852.

Stoutenburg, Adrien. *American Tall Tales*. New York, NY: Viking, 1966.

Tallents, Sir Stephen. *Green Thoughts*. London: Faber & Faber, 1952.

Taloumis, George. "Cutting a lawn gone to meadow," *Boston Sunday Globe*, June 15, 1980, 82.

Thomas, James. "The Seasons." In O. Henry Warren, *The Good Life*. London: Eyre and Spottiswoode, 1946, 237.

Tichauer, E. R., and Gage, Howard. *Ergonomic Principles Basic to Hand Tool Design*. Akron, OH: American Industrial Hygiene Association, 1977.

Tolstoy, Leo. *Anna Karenina* (Constance Garnett, trans.). Indianapolis, IN: Bobbs-Merrill, 1978 (1876).

The Très Riches Heures of Jean, Duke of Berry. New York: George Braziller, 1969.

Turner, Newman. *Fertility Pastures and Cover Crops* (2nd ed.). Pauma Valley, CA: Bargyla and Gylver Rateaver, 1974.

Ucko, Peter J., and G. W. Dimbleby (Eds.). *The Domestication and Exploitation of Plants and Animals.* Chicago, IL: Airline, 1969.

Van Vlack, Lawrence. *Elements of Materials Science.* Reading, MA.: Addison-Wesley Publishing Co., 1959.

Van Wagenen, Jr., Jared. *The Golden Age of Homespun.* Ithaca, NY: Cornell University Press, 1953.

Vaughan, Charles E., Bill R. Gregg, and James C. DeLouche (Eds.). *Seed Processing and Handling.* State College, Mississippi: Seed Technology Laboratory, Mississippi State University, 1968.

Wadsworth, Wallace. *Paul Bunyan and his Great Blue Ox.* Garden City, NY: Doubleday, 1964 (1926).

Warner, Anne S. "On scything," *Country Journal,* August 1975, 96.

Weinsteiger, Richard. "Build-it-yourself grain cradle." In Diana S. Branch (Ed.) *Tools for Homesteaders, Gardeners, and Small-Scale Farmers.* Emmaus, PA: Rodale Press, 1978, 244–245.

White, K. D. *Agricultural Implements of the Roman World.* Cambridge: Cambridge University Press, 1967.

Williams, Christopher. *Craftsmen of Necessity.* New York, NY: Random, 1974.

Worthington, Jim. *Natural Poultry-Keeping.* London: Crosby Lockwood Staples, 1960.

Wright, Philip. *Old Farm Implements.* North Pomfret, VT: David and Charles, 1975.

Zeitlinger, Josef, of Leonstein. *Sensen, Sensenschmiede, und ihre Technik.* Linz-Donau: J. Wimmer, 1944.

Index